Professor Farnsworth's
Explanations in Biology

Also Available from McGraw-Hill

Schaum's Outline Series in Science

Each outline includes basic theory, definitions, and hundreds
of solved problems and supplementary problems with answers.

Current List Includes

Analytical Chemistry
Applied Physics, 2d edition
Biochemistry
College Chemistry, 6th edition
College Physics, 8th edition
Earth Sciences
Genetics, 2d edition
Human Anatomy and Physiology
Lagrangian Dynamics
Modern Physics
Optics
Organic Chemistry
Physics for Engineering and Science
Zoology

Available at Your College Bookstore

Professor Farnsworth's Explanations in Biology

Frank H. Heppner

University of Rhode Island

Illustrated by Cady Goldfield and Kandis Elliot

McGraw-Hill Publishing Company

New York St. Louis San Francisco Auckland Bogotá Caracas
Hamburg Lisbon London Madrid Mexico Milan Montreal New Delhi
Oklahoma City Paris San Juan São Paulo Singapore Sydney Tokyo
Toronto

This book is printed on acid-free paper.

Professor Farnsworth's Explanations in Biology

Copyright © 1990 by McGraw-Hill, Inc. All rights reserved. Printed in the United States of America. Except as permitted under the United States Copyright Act of 1976, no part of this publication may be reproduced or distributed in any form or by any means, or stored in a data base or retrieval system, without the prior written permission of the publisher.

4 5 6 7 8 9 0   DOC/DOC   9 0 9 8 7 6

ISBN 0-07-028351-6

This book was set in Plantin by the College Composition Unit in cooperation
with General Graphic Services, Inc.
The editors were Denise T. Schanck and Margery Luhrs;
the production supervisor was Janelle S. Travers.
The cover was designed by Karen Quigley.
Cover art was done by Cady Goldfield.
R. R. Donnelley & Sons Company was printer and binder.

Library of Congress Cataloging-in-Publication Data

Heppner, Frank H.
Professor Farnsworth's explanations in biology / Frank H. Heppner; illustrated by Cady Goldfield and Kandis Elliot.
    p.   cm.

    Includes index.
    ISBN 0-07-028351-6
    1. Biology.   I. Title.
QH 307.2.H46   1990
574—dc20   89-8301

About the Author

FRANK H. HEPPNER is Honors Professor of Zoology at the University of Rhode Island (URI). A Fulbright fellow, he has also been a President's Faculty fellow at URI and has won many grants and awards for teaching, including a Fund for the Improvement of Postsecondary Education (FIPSE) grant for biology curriculum development. His research is in ornithology, and he is an elected member of the American Ornithologist's Union and a fellow of the Explorer's Club. In addition to 35 research publications, he has written a lab manual, a book on grading and testing for college teachers, and many popular scientific articles. He has long been active in community service and is currently chairman of his local public housing authority. He alternates teaching majors and nonmajors courses in general animal biology, biological photography, and ornithology, in addition to an honors course, "Thinking and Working like a Scientist," which has drawn national attention.

About the Artists

CADY GOLDFIELD is director of publications and public relations for Elderhostel, an international nonprofit educational organization. She has been editorial cartoonist for the Lynn, Massachuetts, *Daily Item,* and her technical illustration credits include work for a book on crabs by the late Dorothy Bliss and a laboratory manual in general zoology. She is a staff instructor at the Museum of Science in Boston and a summer naturalist for the Appalachian Mountain Club.

KANDIS ELLIOT is botanical illustrator for the department of botany at the University of Wisconsin. She has traveled extensively in tropical America on expeditions and is the author-illustrator of many scientific and popular articles. She also illustrates works of science fiction, particularly the stories of Edgar Rice Burroughs, and is a science fiction writer herself. She has degrees in botany, zoology, and applied arts.

Contents

DAY 1
The First Lecture   1

The class finds out how it's going to be.

DAY 2
The Definition of Life   8

The class finds out that life is not what they thought it was.

Study Skills Workshop   26
How to prepare for a science class and take exams.

DAY 3
Simple Chemical Reactions   43

The class learns some relatively painless basic chemistry.

DAY 4
Glycolysis and Fermentation   60

The class dips its toes in energy metabolism.

DAY 5
The Citric Acid Cycle and Electron Transport   76

The class is now up to its neck in energy metabolism.

DAY 14
Evolution and Natural Selection   198

The class gets a little lesson in evolution.

DAY 15
Feedback, Control, and Entropy   213

The class has a dramatic lesson in orderliness.

DAY 16
Nerve Impulses and Countercurrent Systems   230

Professor Farnsworth, fired up, shows how nerves work and fluids flow.

DAY 17
The Biological Basis of Myth and Legend   244

Professor Farnsworth spooks the class.

DAY 18
Reproduction   256

Professor Farnsworth gives a sexy lecture.

DAY 19
Animal Coloration and Patterning   268

The class receives colorful instruction.

DAY 20
Ecology   280

Professor Farnsworth gives a lesson in relationships.

DAY 21
Ethology   293

Professor Farnsworth behaves badly.

DAY 22
The Last Lecture   304

Professor Farnsworth bids farewell.

Foreword from the Editor

What Professor Farnsworth's Explanations Is, and How to Use It

Most students take their introductory biology course in their freshman year of college. The freshman year is often traumatic, but it is never boring. You might be away from home for the first time, having your first experience with group living and meeting new, strange, and perhaps weird people. All these experiences are fascinating, but sooner or later, if you take a biology course, you will be faced with the necessity of somehow dealing with a monstrous textbook, lectures that go by with blinding speed, and examinations that always come before you are ready for them.

I've been teaching freshmen for over 20 years and have had literally tens of thousands of students. A fair number of them have found themselves not doing as well as they had hoped or expected. Very few of these students lacked the brains to succeed. What they did lack was the ability to handle the enormous amount of material that is covered in most introductory biology, zoology, or botany courses. By the time they caught on, it may have been too late and disaster hovered over their heads at the end of the term or semester. Consider the lecture. You might have the most brilliant, stimulating lecturer in the world, but biology is quite technical, and you have to take notes. How do you listen to what the instructor is saying, think about it, *and* transcribe what is being said, especially if you're a slow writer and the instructor is a fast talker? Some students try to tape the lecture, but this only postpones the problem—it is relatively easy to study from a good set of written notes, but very time-inefficient to listen to tapes. Note-taking is an acquired skill, but for some students this skill doesn't fully develop for several semesters.

Textbooks pose problems, too. Nowadays, they are often gigantic—because they have to be. Modern biology is so complex, and has so many facets, that a biology textbook *must* be huge. A thousand pages is not uncommon for a biology major's text, and 1200 is getting to be the new "main line" textbook size. Even a nonmajor's text might be 800 pages long. How do you even *approach* a book like this? How do you avoid being overwhelmed?

That's where *Professor Farnsworth's Explanations* comes in. When I heard that a complete set of transcripts existed of the Introductory Biology lectures of my good friend and colleague, Dr. Stephen Farnsworth, I called Steve and asked him if he was interested in making them available to a wider audience than his own students. He has little patience with the nuts-and-bolts coolie labor of editing, and was just leaving for an extended trip to Borneo when I called. To make a long story short, he agreed to let me prepare them for publication. He wasn't sure when he would return, so I must stress that if there are any errors here, they're mine—Dr. Farnsworth hasn't seen the edited version.

Steve is one of the best explainers I've ever met. He has a way of making abstract things concrete and alive. He's a wonderful speaker, and I've tried to capture, as best I can with the printed word, the kind of magic that was often present in his lectures.

Okay, how do you use *Professor Farnsworth's Explanations?* First, let me tell you how *not* to use it. Under no circumstances should you try to use it as a substitute for an assigned textbook. The "explanations" are designed to be used *with* a text— whichever one your instructor assigns. The text is your source for *facts; Professor Farnsworth's Explanations* is your source for, well, explanations. Secondly, don't think of *Explanations* as a kind of study guide. A study guide provides a review, a summary, and practice questions.

All right, then, if *Professor Farnsworth's Explanations* isn't a text and doesn't work the way a study guide does, what good is it? In any subject, there are facts, and there are ideas. Sometimes the facts get in the way of the ideas. A textbook, because of its very nature, tends to be filled with facts but is shy on examples and explanations. A good lecture can provide a perspective on things and give you a framework that will make sense of the facts. A lecturer's body language and tone of voice can tell you where emphasis should be placed. *Professor Farnsworth's Explanations in Biology* is a transcript, actually more like a script, of a very unusual professor's lectures in introductory biology. Combined with the notes you take from your own instructor's lectures, *Explanations* will give you a variety of viewpoints, and lots of everyday examples of biological principles presented in ordinary language. Once you understand the *general* principles, from your own teacher and *Explanations,* you can go back to the textbook without being swamped by the facts you will need to *completely* understand the principles. I've taken a few editorial liberties in assembling *Explanations.* You will not find every topic here that you might expect to find in a modern biology course. I've left out topics that are relatively straightforward and thus don't require a lot of explaining, for example, characteristics of the various groups of animals and plants. You can use your text alone for those topics. On the other hand, I've included some of Professor Farnsworth's lectures on topics I'm sure you'll not find in *any* text. They're here because they're interesting, bizarre, challenging, outrageous, or some combination of the above. I've included some other aspects of his course that I think you might find interesting or useful. He has a workshop on study skills for biology courses and exam hints that many students find worthwhile.

Every biology course is a little bit different, and whether your instructor assigns

Professor Farnsworth's Explanations or you're reading it on your own, always check with your own instructor if you have questions about the material you read here. He or she is your best and closest authority.

It took a certain amount of convincing on my part to get Professor Farnsworth to agree to have his lectures published. When he finally consented, he asked that he be allowed to say a few words right at the beginning. That statement follows as a preface. As he has not yet returned (at the time of this writing) from Borneo, I anxiously look forward to his reaction to the book itself. I hope you enjoy reading it as much as I enjoyed compiling it.

McGraw-Hill and I would like to thank William E. Dunscombe, William County College; David Fromson, California State University; Robert J. Huskey, University of Virginia; Steven M. Lawton, The Ohio State University; Roy A. Scott, The Ohio State University; and Linda R. Van Thiel, Wayne State University for their reviews of the manuscript.

Frank H. Heppner

Preface from
Professor Farnsworth

Modern biology textbooks are wonderful—beautifully printed, lavishly illustrated, and scientifically precise. Of course, with some of them pushing 1200 pages in length, physical strength to lift the book has become almost as important as reading skill in a first-year biology course. The great virtue of these books, completeness, is also their vice. They are so chock-full of facts, history, scientific authorities, experiments, lines of reasoning, and evidence that it is sometimes hard to figure out what they are really saying, no matter how well written they might be. That problem prompted my old friend and colleague, Frank Heppner, to ask if he could prepare my lecture notes for publication. I was reluctant at first—who would want to look at my notes?—but Frank is persuasive, and I saw little harm in it. So then, this will be a book about "forests"—a view from atop a high hill, from which you can see how the land lies. For information about "trees"—those essential details necessary for a complete and ultimate understanding, you will enter the woods, textbook in hand, ready to chop down great numbers of facts. I hope you will then like studying biology as much as I like teaching it. (I have to warn you about something, though. After many years of acquaintance, I have found, to my delight, that Dr. Heppner is a little, well, eccentric, and I don't know exactly what he's going to do with these notes. I suspect the result will be a mixture of the two of us. I'm the straight arrow, if you get confused.)

Stephen Farnsworth

Aboard PanAm Clipper *Southern Cross*,
en route to Manila

Biography of
Professor Farnsworth

FARNSWORTH, STEPHEN—B.S., University of Chicago; M.S. and Ph.D., Stanford. Associate Professor of Zoology. In his mid-to-late thirties, he is a shade over 6 feet tall. Not married. He has been teaching first-year biology for about 10 years. Student opinion about him is wildly divergent. People tend to love him or hate him. Frequently, students who have just taken his course say he makes impossible demands, asks trick questions, and never considers that students might be taking other courses. His partisans are equally vocal, pointing out that he spends incredible amounts of time helping individual students and that the demands he makes on the students are no harsher than those he places on himself. One of his nicknames is "Dr. Death," supposedly because his exams are reputed to be lethal. Junior and senior students, at least the ones who have survived in one of the biology majors, look back fondly on their experience in his class and attribute at least some of their success to the study habits they were forced to learn in his Biology 100 course. His elective classes are generally filled to capacity, even though they have the reputation for being very time-consuming.

In addition to teaching freshman biology, Professor Farnsworth has an international reputation in his scientific field and frequently travels to developing countries on research trips. After these trips, he is often visited by a man from Washington who carries a big briefcase and wears brown, shiny shoes, lending somewhat of an air of mystery to the professor's travels.

His life outside the college is mostly an unknown to the students, although he occasionally makes reference in lecture to local civic activities and an apparently extensive travel background, including diving in Lake Titicaca and meeting the Dalai Lama of Tibet.

His voice is deep and resonant, and he often doesn't use a microphone in the lecture hall with its 400-student capacity. His handwriting is sharp and angular. When lecturing, he uses broad hand gestures and rarely stays behind the podium. His hair is brown and longish, and he frequently has to sweep it away from his forehead.

Professor Farnsworth's Explanations in Biology

Day 1

The First Lecture

The scene: A large lecture hall at a medium-sized state university. About 400 first-year students are filing into the biology auditorium for the first class of the semester. The lecture hall is designed almost as a theater-in-the-round, with a half-moon-shaped stage and seats wrapped around the stage in a semicircle. Behind the stage is a huge projection screen. The room is painted in bright colors, and the sound-reflecting panels mounted on the ceiling look like a Calder sculpture. At three minutes before the hour, with students still rushing in from their previous classes across campus, the houselights dim, and the strains of George Gershwin's march "Strike Up the Band" burst from the huge stereo speakers mounted high on the walls. The students exchange puzzled glances. Music before a class? Maybe this will be different from high school. At exactly nine o'clock, as the last notes of the march echo around the room, the houselights fade to black and a figure enters from stage right as a spotlight picks out the podium in the center of the stage. As he steps up to the podium, we see that he is dressed in a three-piece, charcoal-gray pin-striped suit, tailored with an English cut. A gold watch chain hangs from the pockets of his vest.

He strides purposefully to the center of the stage and mounts the podium. He looks out over the class for a moment, then begins.

FARNSWORTH: Good morning, ladies and gentlemen. My name is Farnsworth, and this—is Biology 100. On behalf of the department of zoology and our entire teaching staff, I want to welcome you to the class that will be the most difficult, challenging, time-consuming, interesting, frustrating, fascinating, and ultimately rewarding one you will take all semester. What the Parris Island Recruit Training Station is to a U.S. Marine, Biology 100 is to the student of biological sciences. Both are places where you learn the basic skills necessary for survival later, and both are demanding. Biology 100 is the foundation upon which your biological training rests, and just as a house is no stronger than its foundation, your training can be no better than your introductory course. The reason for this is that knowledge in science is acquired sequentially. What do I mean by this? Let's say that there is some principle, call it *principle C*, that you actually use as a practicing physician or marine biologist. Well, to understand this principle, principle C, you have to understand another principle first; let's call it *principle B*. Where do you learn principle B? In an intermediate class. Unfortunately, to understand principle B, you have to un-

1

derstand *principle A* first. And where do you learn principle A? You guessed it, in Biology 100.

Biology 100 is a little unusual in that it is designed both for people who aspire to professional careers in the biological sciences and for those whose interest is as a person and a citizen. At the introductory level, the needs of both groups are similar. The professional needs the basic information necessary to prepare for his or her advanced courses; the citizen needs enough information to make intelligent personal and public decisions involving biology, and in our increasingly technological world a knowledge of biological principles is becoming a necessity.

Consider just some of the issues involving biology which also involve personal decisions. Suppose you have an 85-year-old grandmother in a coma and on life-support systems. The physicians say there is virtually no chance of recovery. Should life support be continued? Should you use pesticides in your garden? Should you be a vegetarian? Should you find out the sex of a child before it's born? Consider also public decisions. Should genetically engineered bacteria be released in the environment? Should animals be used for research? Should germ warfare research be done to develop a possible defense against its use by an enemy? You don't have to be a professional biologist to have a need for biological knowledge.

Bio 100 is a world-class course. This might be a little more important for the preprofessionals among you, but it can't hurt the rest. What do I mean by "world-class course"? Well, it is a fact that some of the careers that many of you are destined for are very desirable, but are also very selective. Everybody wants to be

Jacques Cousteau. Everybody wants to be a physician and drive a Mercedes, or heal the sick, whichever. However, only about one in two of the applicants makes it to med school. About one in three gets into vet school, suggesting that it takes more brains to be a vet than a physician. About one in four gets into the good research grad schools. So, it is very competitive—but here is a radical thought. The competition is not sitting in this room. The people in the class here are not enemies; they are allies and friends, because they can study with you, feed you questions, explain things to you, and give you emotional support. Very few of you will be in direct head-to-head competition with your classmates for the same job or professional school. The competition is sitting right now in a classroom at Cambridge, or the University of Tokyo, or Harvard. There is somebody right this instant at Berkeley who is going to be applying to the same med school as you in four years. And do you know why you're going to whip that person's butt then? Because I'm going to whip yours now. I'm going to work you harder than you even thought you could. You're going to study until your eyeballs bleed.

Sleep is going to be a fond memory. You're going to need toothpicks to keep your eyelids open. And why am I going to put you through all this? Because we both want the same things. You want to be a doctor? I want you to be a good doctor, too, because 15 years from now you might do a liver transplant on me; I don't want you to screw up because you never could remember whether the hepatic artery fed the liver or the kidney, which I'm supposed to be teaching you. But being a good doctor is much more than just memorizing the names of parts. It means figuring things out, and Bio 100 will make you think. The world needs good scientists and technicians, people who can reason things out, not just follow rules from an instruction book. It also needs citizens who can make intelligent, informed choices, so that they're not at the mercy of the media and the politicians. You will learn how to think here, or one of us will die in the effort.

A noticeable chill has fallen over the class. As Farnsworth pauses after his last sentence, a low murmur spreads through the student ranks.

FARNSWORTH: I've given you the bad news. Now for the good news. The good news is that you'll have a tremendous amount of help available to you. We will have help sessions, review sessions, and recitation sessions all semester. I will be available for consultation, as will your laboratory instructors. If you get into trouble, don't panic—see us.

Let me tell you now a little about the mechanics of the course. (*Here follow about 15 minutes of instructions*)...and that is how the exams will work. Now, there's a subject I don't like to talk about much, but it is necessary, and that subject is cheating. If you steal someone's car, you deprive him of his property, but if you steal his ideas or study, you are depriving him of his future. Therefore, in Biology 100, cheating will not be tolerated. I get a bad nervous reaction when I find cheating. My psychiatrist says I have a definite and extreme uncontrollable tendency toward violence when I hear about cheating. You can think of me as the "Dirty Harry" of teachers. (*He pauses and stares out at the class.*) If I find you cheating in *my* course, I will personally, *personally* see to it that the best job you ever find in your life is cleaning toilets in the bus station. (*Long pause.*) Now, then, I hope that didn't upset anybody—I'm sure no one is going to want to cheat here, right? Now, let's talk about something else, eh?

As the semester settles down, for many of you it is going to be very easy to become discouraged. The textbook is humongous, the lectures go by too quickly, the exams are on you before you know it, and you have a million distractions and

pressures. It is very easy to give up. Don't do it. You would be amazed at how many of your fellow students are going through exactly the same thing, and most of them, if they just stick with it, will survive. It doesn't take a great deal of intelligence to do well in Biology 100, but it does take a lot of persistence. Just as runners talk about "the wall," the point where pain and exhaustion build to the point at which they think they can't take another step, you will face the wall a couple of weeks into the course. When you do, just do as the runners do: grit your teeth and say "only a little more." Before you know it, you will be through the wall and the rest will be downhill.

On a similar tack, though I hate to sound like a nag even before the semester starts, I have seen so many people fall victim to this phenomenon, that I just have to mention it. Please, *please,* don't fall behind in your reading. Trust me, unless you're an unusually fast reader, there's just so much material that it is very difficult to catch up, and in some cases it's impossible.

Well, with the overture completed, we are now ready to begin the play we call Biology 100. It will have surprise, tension, uncertainty, and a twist at the end. I love teaching it, and I hope you enjoy taking it. When next we meet, we will start our first "real" lecture and try to determine what life is. (*The houselights come up, and students flock around Professor Farnsworth to make schedule changes, ask questions, etc.*)

Day 2

The Definition of Life

As the students settle into their seats, the houselights dim halfway and "Born to be Wild,"
the theme from the sixties Peter Fonda–Dennis Hopper outlaw motorcycle movie Easy
Rider, *blasts out from the PA speakers. As the song ends, the sound of a motorcycle can*
be heard starting up. Presumably the sound is coming from the tape. However, as the
houselights fade to black, the sound gets louder and a headlight can be seen at the au-
ditorium door. Murmurs of "I don't believe this!" and "What is this?" spring up around
the room. The headlight advances into the room, and the students see that indeed, it is a
motorcycle—a customized Harley-Davidson Low Rider, with ape hanger handlebars and
drag pipe exhausts. The rider does not look like Professor Farnsworth. He is wearing dark
glasses, a headband, Levi's that look as though they haven't been washed since 1978, a
T-shirt that says "OUTTA MY WAY SCUMHEAD!" on one side and "BIKE FREE
OR DIE" on the other. He has on a chrome-studded belt, and a knife handle sticks out
of his engineer boots. He does not look like your average college lecturer. He rides the bike
up a couple of steps to the stage, stops in front of the podium, and turns the engine off.
There is pandemonium in the class—students whistling and cheering. Slowly, and with
proper dude demeanor, he dismounts, pulls a filthy handkerchief out of his back pocket,
and flicks a speck of dust off the upswept back fender. He walks over to the podium and
begins to speak in a gravelly, nasal voice.

BIKER: I got a little problem, and I got ta thinkin' you people might be able ta help
me. See, the thing of it is, I been ridin' motorcycles all my life. In good times I ride
a Harley, and in bad times I ride a rice rocket. As ya can see, these are good times—ol'
George here's a real good bike, and like all Harleys, he's got a personality. He's like
me—he just don't wanna start in the mornin'. But I got a buddy, he's got a other Low
Rider. It mightta been right next to mine on the assembly line, but his starts right up
like a charm every mornin'. Now, I think that's very strange. Ya got two machines,
made from the same blueprints, but they both got different personalities.

I may not be a educated man, but I'm not *stupid*, ya know, and I kinda like to think
about things like this. And I began to think that, okay, a motorcycle is a machine, but
the relationship ya got with your bike is not the same as, say, the one ya got with your
toaster oven, which is a machine, too. Ya don't see nobody goin' around with a tattoo
that says "Toast GE or Die," do ya? So, I started ta think about it, and the more I

8

thought, the more I came ta the conclusion that the relationship between bikers and their bikes is sorta like the one between cowhands and their cuttin' horses.

See, I use ta be out west, so I know about this. Ya work with a horse long enough, and it's like ya become a unit, ya know? Ya don't have ta tell the horse nothin', or even use the reins; the horse just seems ta *know* what ya wanna do, and then does it. And after a while, the cowpuncher gets dependent on the horse, and the horse gets dependent on the rider—neither of 'em can live without the other.

So, I says to myself, "Bingo! That's it!" I been with this bike for four years, and I know all his little strange things. I know exactly where he likes his timin' ta run kinda smooth and lopey, and I got his suspension set up so all I gotta do is lean a little, and we turn left. So I was thinkin' about that for a while, and then it hit me, which is why I'm down here askin' ya to help me out.

Ya see, the thing that jumped out at me was that there were a lot more similarities than differences between the bike and the horse. Oh, yeah, they're made outta different materials, but in terms of what they *do*, it seemed to me that they were the same. Now, the only way that I could explain why the horse could do the things it did was that the horse was alive. Well, if the bike could do the same things, ya know what that meant? It meant that the bike was alive, too.

Now, I know what yer gonna say, and that's what I said, too, at first. "That's a bunch a crap; how can a bike be alive?" But then I thought, well, what *is* the difference between sumpin' that's alive and sumpin' that ain't? I don't know what got me started thinkin' about it then, but see, before I hadda drop out, or more exactly, before I was "invited" ta drop outta high school, I took a biology course.

I think the teacher's name was Miss Gulch or sumpin' dumb like that. But I guess she was a pretty good teacher, 'cause I remembered 'er talkin' about this same thing, and she had a way of separatin' livin' and nonlivin' things, and she had this list a "Characteristics a Livin' Things" that we all hadda memorize. So I tried to remember all a 'em that I could, and when I asked myself if the bike had these characteristics, it just about blew me away, because all the ones I remembered, why, the bike had 'em! And alluva sudden the idea that the bike was alive didn't look like such crap anymore.

Well, ya can bet yer life I wasn't gonna go out and make a fool outta myself in front of all my bros at the Club and tell 'em my bike was alive without doin' some checkin' first. See, if the motorcycle is alive, that explains a lotta things, like about the personality, but what the hell, maybe I forgot a couple a those characteristics a livin' things, and that's why I come down here. I figured that between alla ya, ya must know every one a those characteristics, and I could get it all checked out. Your professa, he was real nice about lettin' me in here, and I didn't hafta stomp him after all.

So lemme tell ya first about the ones I remember, and I'm gonna show ya how the bike has 'em; then you let me know if I forgot any.

FIRST, I remembered that a livin' thing respires. That means it uses oxygen. Well, on the bike, the oxygen comes in here, through the air cleaner, and then he burns it up in his cylinders. No difference between that and metabolizin' food. SECOND, a livin' thing takes in energy from the environment. Well, here's where a bike takes in energy from the environment, right through the gas cap here. Now,

I know what you're gonna say—he don't get the energy himself. But does a baby get its energy by itself? Unh-unh, ya gotta stick sumpin' in its mouth too. THIRD, a livin' thing responds to stimuli. Well, boys and girls, lemme tell ya, this here machine sure do respond to stimuli—I twist his throttle and he goes. I tap his brake and he stops. Just like a horse. So far, so good—there ain't no difference between my Harley and a livin' thing. Now, this is where you people come in. See, my memory started ta let me down about here, and I want ya to help me to come up with the rest. I wanna make sure that nobody down at the Club pulls one a those "characteristics" outta his back pocket when I go down there and tell 'em my bike is alive. So I want ya to sing out any a those characteristics ya can remember.

He leaves the podium and steps out on the stage. A forest of hands wave in front of him.

BIKER: Okay, you.

STUDENT: A living thing evolves.

BIKER: A living thing evolves? Ya ever see a picture of a old Harley? Does it look like a new one? How come it don't look like a new one? Because people liked some features and didn't like others, and the guys who make Harleys picked out the good stuff and dropped the other stuff. So the result was that the design changed over time. If that ain't evolution, I don't know what is. Okay, next.

ANOTHER STUDENT: A living thing thinks.

BIKER: Thinks? Is a tomato plant alive? Does a tomato think? Maybe you got strange tomatoes; they write equations or sumpin'? No, we're talkin' about *universal* characteristic things here, awright? Okay, next.

ANOTHER STUDENT: A living thing can repair itself.

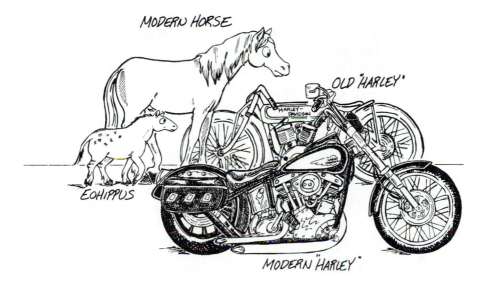

BIKER: Ha! I can tell you ain't never owned a Harley. They fix themselves alla time. But lemme ask ya somethin' first—can a livin' thing fix *anything* that goes wrong with it? Like ya lose an argument with some dirtbag in-a car, and ya lose a leg. Can ya grow back a new one nice and easy? No! Ya can grow back a little skin and fix yourself up, but that's it. Limited self-repair, ya'd hafta call it. Well, it's the same thing with a Harley. They leak alla time—transmission, heads, everythin'. But as soon as ya have a leak, see, the gasket swells up and stops the leak—that leak, anyway, until ya get a new one. Even a car can fix itself a lotta times. Let's say the clamp that holds the distributor gets a little loose so yer timin' is off. Well, the damn engine'll start shakin' so much, that sooner or later, the distributor will vibrate back to the right position, the shakin'll stop, and you'll be okay for a while. So don't gimme no bull about a machine not being able to fix itself. Okay, now yer gonna have to give me somethin' better. What else?

The hands are starting to thin out. The students have the feeling that they are about to be suckered into something, but they don't know quite what. They have been following the biker's arguments closely, but somehow they don't sound right. *It seems obvious; it seems common sense that the motorcycle is not alive, but to* prove *it is not alive appears to be much more difficult than they had thought.*

A STUDENT: A living thing grows.

BIKER: And I suppose this bike existed, just like ya see it here, ever since the beginning of time, eh? Ya know, the Moslems, they go to Mecca, but my friends and me, we go to York, PA, 'cause that's where the Harley assembly line is. And that's where bikes grow. Wait a minute,—okay, you, with the hand waving around.

ANOTHER STUDENT: But that's not really growing. You don't have a little tiny bike growing into a great big bike. You have a frame, and then you add on the engine and all the other parts.

BIKER: Aw right, good point. But is a grown-up human the *same shape* as a baby, only bigger? No way! The adult and baby don't even have the same organs. Some things disappear; others appear. The point is, both the human and the machine get bigger as they get built. Okay, you next. I can see if I don't call on ya soon yer gonna split a gut.

THE STUDENT: (*Very satisfied*) The living thing is made out of CELLS!

BIKER: (*Pauses*) A wise guy, huh? Okay, now you're talkin' about sumpin' that's made outta cells, like a mushroom or a jellyfish, not sumpin' like a amoeba that lives by itself and *is* a cell, izzat right? A single cell is a whole different story, and maybe I can talk about it later. Okay, what is a cell, when ya compare it to the animal or plant that it came outta? It's a subassembly, that's what it is. It has parts of its own, it has a job ta do inside the body, but it can't make it on its own. (*Hands are up all over the auditorium*) I know what yer gonna say! Yer gonna say the cell can live on its own, and that's the difference. But I'm tellin' ya it can't. If I take my sticker here (*reaches down and pulls the knife out of his boot*) and come down and carve a couple a dozen cells offa yer leg, then take 'em up here and put 'em on this little table here, ya know what's gonna happen to 'em? They're gonna die, that's what. They can't get oxygen on their own, they can't get food, they can't get rid a their crap. They're gonna die in minutes, unless I provide their needs like the original body did. A cell can't live outside its body without artificial help. There ain't nothin' unique about a cell outside its body. A carburetor, which is a subassembly, ain't nothin' outside the engine it's part of, and a cell ain't nothin' outside the body it's part of. So when ya tell me that a living thing is made outta cells, all you're tellin' me is that it's made outta subassemblies, and that's what a motorcycle is made outta too. You can do better that that. I thought I was talkin' ta experts here. Any more ideas?

A STUDENT: A living thing reproduces.

BIKER: Oh, boy, ya know I been thinkin' about that one. That's tough—yer right. Like, if I park my bike inna garage next to a big police bike overnight, they're not gonna have a litter of little mopeds six months from now. So maybe that's it, eh? Whaddya think, is that maybe the only real difference between a livin' thing and a nonlivin' thing? The livin' thing reproduces, and the nonlivin' thing don't. How

many of ya think that's it? Then maybe we can settle this, and I can go home. Raise hands, how many of ya think that's the difference?

Many hands raise, but many hold back, suspecting a trap. They are probably right.

BIKER: Gee, a lotta ya think so. Well, ya know, if that's really true, that makes me real sad. Ya know why? 'Cause I'm gonna hafta go right down to the poolhall and tell my grandmother that she's not alive 'cause she ain't never gonna reproduce.

Consternation in the class. Shouts, laughter, students jumping up and down in their seats. They've been had, and they know it.

BIKER: Oh, yer tryin' ta tell me now that it's not whether a *individual* reproduces, it's whether the *kind a thing it is* reproduces that makes it alive instead a nonlivin'. Is that it? It's the *potential* ta reproduce that makes sumpin' alive? Awright, let's try it again, so I can get outta here. How many a ya think the potential ta reproduce separates livin' from nonlivin' things?

There is much hesitation. This sounds right; they can't find a flaw with it, but after being caught before, they're reluctant to commit themselves.

BIKER: C'mon, ya wimps, what's it gonna be? Yer gonna have ta decide. Anybody what doesn't vote, I'm gonna come out there an' trash ya. Try it again. How many think the potential ta reproduce defines life?

Reluctantly, most hands raise.

BIKER: How many think it don't?

Practically no hands go up.

BIKER: (*He rubs his hands together gleefully, then reaches in his back pocket and pulls out a roll of bank notes, wrapped on the outside with a hundred dollar bill.*) That's great, that's great, that's what I thought ya were gonna say. Now, I have $10,000 here that I'm gonna give to anybody who can produce for me the offspring of a mule. Mules are sterile at birth. No baby mules from a mule. Sterile. The only way ya get a mule is to breed a horse and donkey, and the offspring that ya get is sterile. So here ya got a animal that walks, runs, craps, and can pull a plow, but it AIN'T ALIVE 'CAUSE IT CAN'T REPRODUCE! Waddya think about that, pencil necks?

The class is in an uproar. Yells and shouts erupt from the students.

BIKER: Awright, awright, calm down a little. Ya been payin' good attention here, so I'm gonna straighten ya out and tell ya what this all means.

He ducks behind the podium. Dark glasses and headband come off, hair is slicked back, and Professor Farnsworth stands up. With these few simple changes, the transformation is remarkable. When he speaks, the sing-song is gone, and the voice is a half-octave lower.

FARNSWORTH: What it means is that there is no single definition that will separate all living from all nonliving things. I must say, you did very well, though. Some of

your suggestions were excellent. I was worried that you would come up with the one that I was most afraid of—that the living thing transmits the information necessary for its construction from one generation to the next. If we had been talking about machines in general, that would have been no problem, but it would have stumped me on the bike.

Well, what *does* it all mean? This is a biology course, so it's appropriate to talk about what *biology* is—it's the study of living things, naturally. Our biker buddy might have terrible language, but he at least has the merit of thinking seriously about a problem that has plagued scientists, philosophers, and people in general for centuries. Biology is the study of living things, to be sure, but we start getting into trouble when we start to ask what a living thing is—if we are very fussy and want to have a definition that will include all things which are "living" and exclude all things which are "nonliving." The problem is more complicated now than it ever has been.

First, there is the natural problem of the viruses. A *virus* is a submicroscopic entity, really a collection of molecules, that has none of the "traditional" properties of life, like respiration, responsiveness, and so on, but it *does* reproduce. However, to make our lives miserable, these infernal things do not reproduce by themselves— they invade a cell which is living by any definition and screw up the cell's molecular manufacturing mechanism so that it starts making viruses. So the virus does not actually reproduce itself, but it *becomes* reproduced—a subtle but important difference.

The argument our biker friend raised about the sterile hybrids has merit, but it is essentially an artificial situation—the parents of the sterile hybrid would normally never mate in nature. Nevertheless, the existence of the sterile hybrid prevents us from having a strict definition of life based on reproduction.

The development of modern technology has raised some intriguing challenges to our traditional definition of life, which sooner or later has reproduction at its core. Let me ask you this. Do you think it would be possible to build a self-reproducing machine, and if so, how would you do it? (*He pauses to let this thought sink in.*)

Well, the answer would appear to be "Yes, it would be possible." A human, or group of humans, would have to build and set up the first one, but after that it could go on its own. The key to such a self-reproducing machine would be a computer with a program that could make copies of itself. Most of you by now know that a computer program is just a set of instructions that tells the computer what to do. One of the instructions might be to make a copy of the set of instructions. Another instruction might be to make a new computer from parts assembled by a robot directed by the computer. Still another set of instructions could be directions on how to build new robots that the copy of the program could use to build a new computer.

This self-reproducing machine would need raw materials from someplace. It would need silicon, and gold, and steel, and plastic—as a dog needs carbon, nitrogen, and oxygen. How does the dog get its raw materials? It takes them from the environment. The dog is dependent on its surroundings to live and reproduce. Our self-reproducing machine would also be dependent on its environment, but it might acquire the raw materials in a different way. It would buy them. We would equip our very first machine with a lot of solar cells, which the machine could use to make electricity, which it would sell to the power company. With those funds it would buy silicon, which its robots would make into new solar cells for the new self-

reproducing machines. So our first self-reproducing machine might be like a virus—capable of reproduction, but not of independent reproduction.

A truly independent self-reproducing machine would be enormous because it would have to have robot bulldozers and robot mines for its steel, robot airplanes and ships to bring its raw materials in from overseas, and buildings to house all its machinery. But there is no theoretical reason why such a machine could not be built.

If we want to make a living thing, though, making a machine like this seems like the hard way to go about it. How about if we duplicated, in the laboratory, the kind of life we are familiar with—a life based on chemicals and chemical reactions. Test-tube life, or synthetic life, in other words.

Such a thing is theoretically possible, but we are quite a way from it now. What do we actually mean by saying "test-tube life"? Well, we know that the chemical essence of life as we know it here on earth is a very large molecule called *DNA*, which we'll talk about at great length later. To say truly that you have manufactured life, you would have to construct such a molecule out of raw materials, and then have it reproduce itself and perform the functions of natural DNA. Molecular biologists are close to this, but are not there yet. Again, like the self-reproducing machine, there is no theoretical reason why this couldn't be done.

Now, why have we spent so much time talking about things that admittedly are much closer to science fiction than science right now? Because, one, science has a way of advancing much faster than we think it will, so it's good at least to consider what appear to be fairly remote possibilities, and, two, thinking about what artificial life might be like focuses our thoughts about natural life.

All right, we've seen that it is tough to tell the difference between something that's alive and something that is nonliving. I have another one for you. What is the difference between something alive and something dead? C'mon, don't just sit there. Give me something. Anything.

The students are a little reluctant at first, but after a few seconds, some of the bolder ones tentatively raise their hands.

FARNSWORTH: Okay, in the red shirt in the middle. What do you think?

STUDENT: You're dead when your heart stops beating.

FARNSWORTH: Excellent! How many of you think that's right? A lot. Keep that in mind while I ask the next question. What do you call it when you deliberately make somebody dead. Just shout it out.

THE CLASS: Murder!

FARNSWORTH: Wonderful! This is going to be easier than I thought. There's one little thing that bothers me. You know open-heart surgery? Where they repair the valves and sew up rips—things like that? Well, the surgeon has to stop the heart in order to work on it. So, if you're dead when your heart stops beating, and the surgeon stops your heart, and somebody who makes you dead is a murderer, then I guess we're going to have to arrest all the heart surgeons, right?

Confusion. This can't be right, but why it's not right is not immediately clear.

FARNSWORTH: And, you know, there's another thing that bothers me just a little. Turn that statement around. If you're dead when your heart stops beating, that must mean that if your heart is beating, you're alive, right? But what about all the other parts? Modern medicine is wonderful. We can keep that piece of heart meat going almost indefinitely. Everything else—brain, kidney, nerves—could be shut down, but we could keep that ol' ticker ticking away. Is somebody in that condition alive? What if we removed the heart, kept it supplied with nutrients, stimulated it to keep it beating, and put it in a bottle, while we buried the rest of the body? Would the person still be alive? Could the "deceased's" family collect on the per-

son's insurance, so long as that heart was beating? Is anybody else bothered by this? Yes—the Rams T-shirt on the left.

STUDENT: We can't use heartbeat as a definition of death any more because of medical technology. Now they use brain death.

FARNSWORTH: Ah, brain death. That would be good. But how do you measure brain death? I've had students I thought were brain dead after an exam. Can you tell if somebody is brain dead just by looking at them? You can tell if somebody's heart is going pretty easily—just look for a pulse. But what if they're just unconscious. Or in a coma. What's the difference between somebody in a coma and somebody who's brain dead? Well, usually, you have to hook somebody up to a machine, an electroencephalograph, to tell if they're brain dead. What state are you in until they hook you up to such a machine? Let me give you an example.

A husband and wife. "Green" is their name. In the husband's will, he says that if he dies before his wife, his estate goes to her. If the wife dies before he does, the estate goes to the college here. And we're talking big bucks, ten million smackers. But the wife has a will too. In her will, all she says is that if she dies, her estate goes to her son by a previous marriage, a greedy little roach named Freddy. Needless to say, both the president of the college and Freddy have a very lively interest in the state of health of Mr. and Mrs. Green. Now, here's a lonely country road. Patch of ice on the bridge. The Greens' car skids out of control and hits a telephone pole. Yuck! A really *ugly* accident. Mr. and Mrs. Green are smooshed all over the road. A doctor arrives at the scene and takes Mrs. Green's pulse first. No heartbeat. Then Mr. Green's. No heartbeat. But, they have a CPR guy in the ambulance, and he tries to get them both going on the way to the hospital. The president of the college and Freddy both hear about the accident on their scanners and race to the hospital. By the time the ambulance gets to the emergency room, the CPR has stimulated a couple of beats in Mr. Green's heart, but has done nothing for Mrs. Green. The doctor sets up the electroencephalograph and hooks up Mr. Green first. Nothing. Flat as a board. Brain dead. Then he rolls the cart over to Mrs. Green. Same thing. The doc shakes his head and says, "How sad, how sad. I pronounce them both dead." In the meantime, Freddy and the president are both jumping up and down, screaming, "Which one!? Which one!? Which one died *first!?*" The college's lawyer is yelling "Mrs. Green! Mrs. Green! The doctor measured her heartbeat before Mr. Green's." Freddy's attorney is ranting "Mr. Green! Mr. Green! The doctor measured his brain waves first!" After a couple of months, the matter comes to court, and you students are the jury. Who are you going to give the money to? Freddy or the college?

Animated discussions pop up all over the room. Farnsworth lets them continue for a minute or so, then continues.

FARNSWORTH: Okay, okay, you get the idea. Here is something that you're used to thinking of in black-and-white terms—you're alive or you're dead, and now we find it is not so easy. And we have hardly even begun to touch the questions. Brain dead, heart dead, what difference does it make when you're talking about a lettuce

plant? When is lettuce dead? For that matter, when is an amoeba dead, for heaven's sake? How about a seed? When is a seed dead? And suppose we had a computer that could think. If somebody pulled the plug on it so that it lost its programs and its memory, would that person be guilty of violating the computer's civil rights? Doesn't it stand to reason that something that can think would have civil rights? But if thinking is the basis for civil rights, does that mean people in comas have no civil rights because they can't think?

It is very clear that science has gone much faster than our common law has. There isn't any agreement about how to define when death has occurred, as we've seen. There certainly isn't general agreement about when life begins—that's the whole abortion controversy. And now when you plug in the whole matter of advanced medical methods in human reproduction, such as frozen embryos, things get real messy real fast. There was a case in which a woman had some of her eggs medically removed from her body and then fertilized in a test tube with her husband's sperm. The resultant embryos were frozen, and the idea was that one of them would be reimplanted in the woman's body. But before that could happen, the couple was killed in a plane crash. Who decides what happens to the frozen embryos? Can you flush them down the drain? Can you sell them to a couple who wants children? You can't sell a baby legally, and you can't even give it away without a lot of legal rigmarole. How about an embryo? Could you do medical experiments with it? These are all unanswerable questions that sooner or later will have to be answered. And this is a good time to introduce you to an important idea about questions. To show you what I mean, I need somebody who has a boyfriend or a girlfriend. Anybody. C'mon, don't be bashful. Ah! Okay, ma'am, you have a boyfriend, correct? And how long have you been going out with him? Six months. Fine. Now, could you tell me, please, yes or no, have you stopped beating him up yet?

STUDENT: (*Outraged*) No! Yes! I mean—that's not a fair question.

FARNSWORTH: Exactly. But it is a specific kind of question. The Zen Buddhists would call that a *mu* question. A question that has no answer, because it carries with it an inherent assumption which may be false. In this case, the question carried with it the assumption that you were trashing your boyfriend. That assumption may or may not be true, and before you could deal with the question I asked, you would have to take a step back and see first if there was an assumption in the question and, secondly, if that assumption was true. Another question. In what exact year did Columbus discover that the world was flat? Give me the exact date. No? What's the problem? Ah! Very perceptive—this is a *mu* question because the assumption is that Columbus discovered the world was flat, and that assumption is false.

Now, let me ask you this. What is the difference between something that is alive and something that is not alive? Aha! Maybe this is a *mu* question. Is there an assumption here? Of course. And what is that assumption? That there is, in fact, a definable, sharp, distinct difference between living and nonliving. Maybe that assumption is true; maybe it isn't. Some things, certainly, have a distinct line. Can you be a *little bit* pregnant? Of course not—you are or you aren't. Can you be a little bit alive, or mostly dead, or nearly living? As we've seen, from a biological standpoint, the answer is "yes," you can be somewhere in between life or death, or somewhere in between living or nonliving. So why does it make us so uncomfort-

able when we hear that? Because so many human activities *require* a sharp division between things, whether they really exist or not. We've seen that the law *has to have* a fixed time of death, exact to the second, even though, as biologists, we know that there is no such thing. Many religions, too, need an exact time of death, and an exact time of beginning of life, to determine when the soul leaves or enters the body. Let me give you an example of what I'm saying—about beginnings and endings. (*He steps out to the middle of the stage and holds his arms out, with both elbows bent and fingers touching.*) What kind of geometric figure am I making here with my arms? That's right, a circle. Now, can somebody tell me where the beginning of the circle is? (*Confusion reigns.*)

STUDENT: There isn't any. A circle has no beginning or end.

FARNSWORTH: No, no, no. Don't you see it? See; the circle begins right here where my watch is, goes all the way around, and then ends up here on the other side of the watch. (*Slaps his forehead*) But, oh, how inconsiderate of me. Those of you over on the right can't see my watch from where you're sitting. For you, I make a special deal. For you, the circle starts up here, where I have the pack of cigarettes rolled up my sleeve, goes around past the watch, you can't see it, take my word for it, it's there, then winds up here on the other side of the pack. Now, is everybody happy with that? (*Hands waving all over*) No? What's the matter?

STUDENT: The circle really doesn't have a beginning or end. You just put a beginning on top of it that was convenient for you, and when that wasn't convenient for the people on the other side of the room, you moved it. The real circle doesn't have a beginning or end, no matter what you put on top of it.

FARNSWORTH: (*His eyes bug out.*) What a class I have! I'm going to go right over and kiss the dean of admissions. You've hit it exactly right. The *real* circle doesn't have an end, but we stick one on because we need one. So long as that end is

convenient, we'll keep it. If it becomes inconvenient, we'll move it. So life really doesn't have a sharp end, but so long as heartbeat is convenient, we'll use it. When heartbeat becomes inconvenient owing to technology, we can change our definition of death. The *fact* is that life *faaaaaades* into death. Nonliving *faaaaaaades* into living. A motorcycle is more alive than a toothbrush. A robot is more alive than a motorcycle. A dog is more alive than a robot. When something gradually fades into something else, that's called a *continuum,* and in biology we have to get used to continua because there are so many of them. We also have to get used to *mu* questions because just as our unfortunate friend over here got tricked on the question about her boyfriend, nature can trick us with questions that have hidden false assumptions.

Let me see how if I can tie this all together somehow. This is a course about biology, biology is about life, but life—or death—cannot be precisely defined in such a way that one definition applies to all situations, plant, animal, or robot. Modern technology is blurring the old sharp lines between living, dead, and nonliving, and the law is going crazy trying to deal with situations that were unimaginable only a few years ago. As people who will have had training in biology, you have a special responsibility to think about these things, because they seem to go against common sense, and your opinions will be listened to by people who have not had the benefit of our training.

Now, because you have been so attentive today, I have a present for you. Tomorrow night, at 7:30 in room 237, I will have a study skills workshop for biology courses. You don't have to come if you don't want to—I won't take attendance—but it is kind of silly to throw away an opportunity like this. Be prepared to spend a couple of hours.

If any of you now still have lab section changes you need arranged, come on up and I'll take care of it for you. And remember tomorrow night.

About 20 students come forward, while the rest scramble out to their next class.

Study Skills Workshop

A room smaller than the auditorium. About 50 students are sprawled over the desks, sluggish from trying to digest the dining hall supper. At precisely 7:30, Professor Farnsworth appears. He is wearing a shirt that looks vaguely African, and he's gnawing on the end of a chicken drumstick that clearly has seen better days.

FARNSWORTH: Well, you must be all the A students. (*Hopeful laughter*) It's true—it seems as if the only people who ever come to these things are the ones who don't need them. I wish I could promise you that what I will tell you tonight will really give all of you A's, but I can't do that. All I *can* promise is that if you more or less follow these suggestions, you will do as well as you can.

The fact is that not everybody starts this race from the same starting point. Some of you went to rotten inner-city schools, and some of you have come from good prep

schools. Some of your parents are Ph.D.s, and others haven't opened a book since they left school. The ones I feel sorriest for are the ones who did very well with little effort in high schools that didn't make many demands on them. They face the greatest danger, because by the time they realize the rules have changed, it sometimes is too late.

However, just because somebody comes in with an advantage, that doesn't necessarily mean he or she hangs on to it. We have had tough ghetto kids come in here, who knew what they wanted, and they simply blew the preppies out of the water. So what it comes down to is this: whether you've had the advantages or not, everybody has a shot at it.

Okay, study skills. Every course is a little bit different.* You don't prepare for an English course in quite the same way you do for a chem course, and two different instructors for the *same* course might have dissimilar expectations, so your first task is to find out, as best you can, what those expectations are. The place to start is the course syllabus, if there is one. At a minimum, this will tell you what the lecture topics are, the dates of the exams and assignments, and the readings. Some course outlines will contain a lot more useful information. Read and heed this outline. It is absolutely amazing to me how often students just throw away their good grades because they don't read instructions. I've always tried to make my instructions on exams as clear as I could make them, but still, a lot of students didn't seem to follow them and would do things that would unnecessarily cost them points. Were they being nonconformists? Were they rebelling against formal instructions? I didn't know, so I decided to try an experiment once. I made up an exam question—a real killer. This question was so hard it would have made Einstein cry. It took 15 minutes just to read the question. I put this question on an exam; it was question 7. The instructions to the exam were printed on the answer sheet, and they went something like "Use only capital block letters. There is only one correct answer to each question. The answer to question 7 is 2,713. If you think this is a trick, raise your hand and I will confirm that that is the answer." So after the exams were distributed, and the students had them in their hands, I said, "Now, before you start, be sure, absolutely sure, to read the instructions for the exam, because they are very important." Now, what fraction of the class do you think got question 7 wrong, with the answer printed right there in the instructions on the answer sheet? You won't believe it, and neither did I—40 percent of the class got it wrong. The lesson is don't throw away anything that is given to you free. Read the syllabus and any instructions. Save them in your course notebook for future reference.

Now, you have more courses than just Bio 100, and you have only so much time during a week. How do you decide how much time you can afford to spend on Bio 100? Well, a rough rule of thumb for figuring out how much study time is expected in a semester-long science course is: two hours outside of class every week for every

*Editor's note: Professor Farnsworth's suggestions and recommendations are based on *his* course. Although much of this material will be of general use, be sure to check with your own instructor for specific details and requirements.

credit hour of the class. Bio 100 is a four-credit class, so eight hours a week, on average, is a reasonable figure. That eight hours doesn't include taking coffee breaks, checking out the guys at the next library table, or staring into space to compose your thoughts. So, you figure, if you're carrying 16 credits, we own your brain for 32 hours a week outside of class time. If you have a fair number of labs, 16 credits is probably about 23 hours in a classroom, so we have you for 55 hours total a week. Since there are 168 hours in a week, figure 7 hours a night for sleep, that leaves you with 64 hours a week to eat, work, play, or fool around. Not too bad.

So far, however, we have been talking about an *average* work week. There is really no such thing as an average week, though. Academic life is very erratic. Papers due, exams, things seem to bunch up so that some weeks are very quiet and others are horrendous. This is where your first important tool comes in. Get yourself a calendar, one of those big kinds that has a space for notes for each date. On this calendar, write all the dates when you will have quizzes and papers and exams. When you do this, you will immediately see that in some weeks there will be a pileup of work. So much so, that you probably would not be able to get everything done if you waited until the end to do it. What does this tell you? That you must do some things way *ahead* of when they're due. This is a big difference between most high schools and here. In high school, you really could get away with putting everything off until a couple of days before the due date. Also, high school teachers

How Much Time for BIO 100?

① 1 credit hr/class ⟹ 2 outside study hrs/wk
② Bio 100 = 4 credits ⟹ 8 hrs/wk study time
③ 16 credits ⟹ 32 hrs/wk study

So: 16 credits + Lab = 23 hrs in class/wk
                        + 32 hrs study/wk
                          55 hrs/week WORK

⚃ 168 hrs/wk (absolute)
  − 7 hrs/night × 7 (sleep)
= 119 hrs/week conscious (EAT-WORK-PLAY)

(**) − hypothetical average

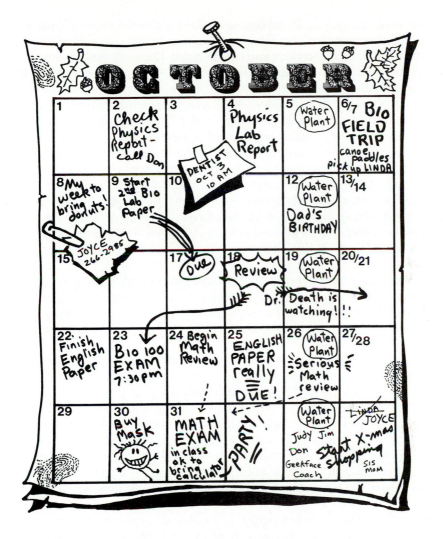

tend to remind you constantly about upcoming assignments and tests. Those days are gone forever, and the calendar will serve as a replacement for all those human reminders.

In figuring out a calendar, you will want to include padding time for emergencies. Suppose this is September 27, and you know that you have a paper that will take you a week to do and is due on November 20. When do you start the paper? November 13? No, because maybe it will take you a little longer than estimated. November 11? What happens if you get the flu for three days on the sixteenth? A rough rule of thumb in figuring when to start a project that has a fixed deadline is to add 25 percent to your time estimate if you have never done this kind of work before, 10 percent if you have, then add another 10 percent to allow for unforeseen occurrences. So, if you figure it will take you 2 weeks to do a semester paper, and you've never done one before, 25 percent of 14 days is about 4 days, making 18 days. Add your 10 percent for emergencies, or another 2 days, and you now have 20 days to do a project that you think you can do in 14. The more projects you do, the better will be your estimating ability, but at first be very, very conservative. It also helps to set a personal deadline for a project a couple of days ahead of the real deadline. Much better to finish a couple of days early, because many instructors will not accept late work *at all*, regardless of your excuse. Why take that chance?

Now we have a semester calendar, but we need to make another kind of calendar, a sample weekly calendar. Let me ask you this—how many of you are night people, doing your best and most concentrated work after nine o'clock? Okay, quite a few. How many are morning people—get up real early and get reading done before breakfast? Again, a fair number. On another tack—how many of you can get good studying done if you have an hour between classes? Some. How many of you need an uninterrupted couple of hours? A lot more. Okay, the point is this. Everybody has a different biological study clock, and this clock must be accommodated when planning your study day.

For example, somebody who needs a big block of time to establish concentration shouldn't try to figure on any studying between classes. That will be the time for personal errands, talking with friends, etc. Somebody who can use that time profitably might be wise to bring a text or notes to class to study in the hour before the next class.

The first thing you do, then, in making up your weekly calendar, which will go from the time you get up to the time you go to bed, will be to plug in your fixed time obligations. Classes, sports, labs, part-time jobs, family responsibilities. Then add meals and personal necessities like showers. What you will be left with is a series of "holes" on your schedule, into which you will insert your blocks of study time. In general, most people find it better to space their studying on a particular subject through the week. In other words, you wouldn't put in all eight hours of Bio 100 time at one stretch. An hour and a half per subject is a pretty good rule, because that will let you take one day of the weekend off and at least one night for R and R.

Your final calendar might look like this. (*He puts a sample calendar on the overhead projector.*) You'll see that it has some vacant places. That's fine; nobody can be productive all the time—you need some time-wasting time to let your brain rest. You'll start to get into trouble when the time-wasting time starts to exceed the productive time.

Having made this calendar, you will not take it too seriously. You haven't included unavoidable disasters and special events. This calendar is a guide only, to give you a general idea about how your week should run. Without having gone through the exercise, however, you won't have a good feeling for how you should arrange the week.

Now we're ready to talk about studying itself. In a science course, you have two main tasks. First, you have to master and memorize a fairly substantial vocabulary, and secondly, you have to come to an understanding of principles and concepts, some of which will be rather complicated. The way you deal with both tasks is

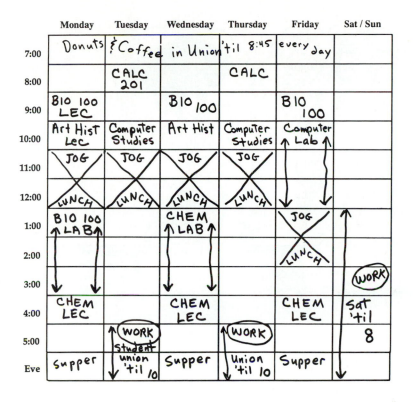

	Monday	Tuesday	Wednesday	Thursday	Friday	Sat / Sun
7:00	Donuts	& Coffee in Union 'til 8:45 every day				
8:00		CALC 201		CALC		
9:00	BIO 100 LEC		BIO 100		BIO 100	
10:00	Art Hist Lec	Computer Studies	Art Hist	Computer Studies	Computer Lab	
11:00	JOG	JOG	JOG	JOG		
12:00	LUNCH	LUNCH	LUNCH	LUNCH		
1:00	BIO 100 LAB		CHEM LAB		JOG	WORK
2:00					LUNCH	
3:00						
4:00	CHEM LEC		CHEM LEC		CHEM LEC	Sat 'til 8
5:00		WORK Student		WORK		
Eve	Supper	union 'til 10	Supper	Union 'til 10	Supper	

they should just try to memorize the whole book and all the lectures as best they can. That may have been a productive approach in high school, but it won't work here. It won't work, first, because there is simply too much material to memorize. It won't work, second, because we're not going to ask you just to parrot back the book. You'll have to parrot back the vocabulary, but we'll ask you to solve problems using the concepts and principles, and if all you've done is memorized some phrases in the book, you'll be dead meat on an exam.

So, how do you prepare for this kind of challenge? Just read the text over and over, or maybe highlight with a yellow pen? No, I suggest you use something called *active studying*. It's a lot of work, and it isn't necessarily the best approach for everybody, but it's a good starting place.

The first part of active studying is development of a good attitude. The student who knows exactly what he or she wants out of life has it easier than the one who is undecided, because the committed student *knows* that the material is relevant to the future, and this certainty makes the studying less work. If you can convince yourself that the material in the course will be useful to you, as preparation for the practical things you will learn in your advanced courses, your study will be a lot more enjoyable. If you don't understand why you are being tormented with all this garbage, well, of course it will be hard for you to develop any enthusiasm. Well, what if you *are* uncommitted to a particular direction? There is at least a *potential*

that the material will be useful for you, and until you do make up your mind, why close off any possibilities?

All right, now we have a good attitude—what's next? Let's start with a text assignment. Chapter 23, let's say. First, I'm going to suggest that you *not* read the chapter summary, even after you've finished the chapter. You're going to save that chapter summary for something valuable later. The first thing you will do is read the chapter over quickly, almost like a novel. I mean very quickly, a skim, really. As you do this, you have a task in mind. That task is to explain to an imaginary 12-year-old kid what the chapter is about, in ordinary, simple English, after you've finished your first reading. Do you see the purpose of this? You want to get a general idea, an overview, before you start your detailed study, and it is very easy to get bogged down in facts. So skim that baby the first time through and then afterwards actually explain, talking out loud, what the chapter was about to the imaginary kid, or another student in the class. The best way to learn something is to explain it to somebody else.

The *second* time through the chapter, you have two tasks—to get a handle on the vocabulary and to identify those ideas that you don't grasp right away. The vocabulary problem can be addressed with flash cards or cue cards. Get yourself a stack, a *biiiig* stack, of 4 × 6 file cards. As you go through the chapter, every time you see a word you don't know, or a word in boldface or italics, write it out on one side of the card, saying the word aloud as you write—this is very important. On the other side of the card, write the definition. This does two things for you. It gives you a handy review aid. While you're lying around, you can thumb through the cards. If you have a boring date, you can pull the cards out of your pocket and send a message. Even more important than *having* these cards is *making* them. There has been all sorts of research

on memorization that shows that the more senses you use in memorizing something, the better that something sticks. Writing it out uses your sense of touch. Saying the word aloud involves speaking and hearing. That's why a homemade set of cards would be much better than a set you bought that was already printed up.

There's another thing about memorization. It's context-sensitive. What the hell does that mean? It means that you can recall something better if the conditions where you are asked to do the recalling are similar to the ones where you did the memorizing. So, if you like to listen to loud rock when you study, unless they play loud rock during the exam, you won't do as well as you would have if you studied under quieter conditions.

Okay, so now you're reading along, making up your vocabulary cards. That's not all you're doing, though. You are reading the text to see if there are any statements you don't understand—are there things that just don't make sense to you? Any statement that you don't understand, put a big check mark by it. If a quick reread of the statement doesn't straighten it out, just go on. Don't try to read it over and over, hoping that something might click. Probably won't happen. *After* you've finished the chapter for the second time, you will have, in effect, a list of statements that need further explanation. The fact that you didn't get the material the first time through isn't necessarily a bad reflection on either you *or* the book. I've found over the years that people are very different in the way they look at things. For some people, the book might have the perfect explanation for a particular subject. For other people, equally bright, the explanation won't make sense, but if it were re-phrased just a little bit, it would be perfect. It's almost like taste in food. Not everybody likes squid soup as much as I do, but that doesn't mean they're bad people. So, not every explanation is good for everybody, and if the book's explanation doesn't do it for you, why then you just have to find another one.

There are a number of ways to do this. Many times, the lecture will cover the same material in the text, from a different viewpoint. You could ask somebody else who's taking the course—although this is a little dangerous. You can't be really sure another person's explanation is a correct one. You could ask a teaching assistant. You can certainly ask me—I'm delighted to talk to students one on one. Just use a little initiative, and you can always find a second explanation.

There's a right way and a wrong way to ask for help, though. If you come to me and say, "I don't understand any of this genetics stuff," there's not much I can do for you. If, on the other hand, you say, "I don't understand in the Hardy-Weinberg equation why it's $2pq$. Where does the '2' come from?" then I can help you right away. It's a two-way street. To get maximum benefit from help, you have to do your homework *first*. Nobody's going to spoon-feed you, but you can get all the legitimate help you need.

Okay, now you have your questions answered, and you're on top of the vocabulary. It's time to go over the chapter a third time, and this time you will be able to go through it fairly quickly. Not a skim, really, but by this time you have the vocabulary solid, and you don't have any unanswered questions. This will be your chance to consolidate your knowledge.

After this third reading, you are ready for your first knowledge test. Close your book, and write out a two- or three-page chapter summary from memory, then check it against the book's summary. Do you see now why you didn't look at the book summary the first time? You might have had a tendency to try and memorize it, which is exactly what I *don't* want you to do. If you have truly mastered the chapter, you won't have any particular trouble writing out your own summary in your own words. You are now ready for the next and final step of your chapter study.

If you did okay in your summary writing, find a friend. Ideally, the friend should be taking the course with you, but that isn't even really necessary. Give your friend the textbook, point out the proper chapter, and have him or her fire questions at you. Not just a few questions—take at least a half hour for each chapter. Obviously, a friend who is taking the course will be able to ask more sophisticated questions

of you than somebody who doesn't know anything about biology, but having a textbook in your lap is a great equalizer.

It is very important that this questioning be done well before an exam. If you do it just before the exam, there will inevitably be questions you can't answer. With an exam looking at you the following morning, all that will happen is that you will become panicked. The night before the exam, go to a movie. In high school, it was possible to pull an all-nighter and do okay, and that is even possible in some college courses, but it won't work here. You need to be mentally alert for the exam, and you won't be if you've stayed up 'til five in the morning.

After this question-and-answer session, you will know two things. You will know what you know and what you don't know. The things you know you won't have to bother with anymore, except in review. The things you don't know you can go back and restudy.

That, then, is the basic approach to studying a chapter. Any questions now?

STUDENT: How about using a study guide?

FARNSWORTH: That depends. If your instructor *assigns* a study guide, it would be foolish not to use it, because you can be fairly sure that the instructor is going to base some of the exam questions on the guide. If it is *not* assigned, then it gets a little tricky. You don't want to use a study guide *instead* of doing the text reading, so if you're a real slow reader, the guide might be counterproductive—you'll need all the time you can get on the book. Also, the study guides rarely have really challenging questions, so the potential A student might be better off with supplemental readings like *Scientific American* reprints. But for the great middle group, a guide does have potential benefit. Okay, next question?

STUDENT: What about highlighting?

FARNSWORTH: Well, I suppose for some people, highlighting or underlining might be helpful, but I have found that what that encourages you to do is memorize phrases without trying to understand them. Let me give you an example. Most of you took high school biology. Let me ask you a simple question. What does the kidney do? Anybody?

STUDENT: It filters the blood.

FARNSWORTH: Good, now, what does it remove from the blood?

SAME STUDENT: Impurities.

FARNSWORTH: Fine, but where do these impurities come from? Impure water? Impure air? Would a frog that lived in pure spring water, and breathed smog-free air, and ate only organically grown flies need a kidney?

SAME STUDENT: No, well maybe they're not impurities like that, they're, like, body wastes, you know?

FARNSWORTH: Well, in your food, what do you waste? Why bother to eat something you're going to waste? And anyway, I thought the waste material went out of the body in the form of feces. So what are these supposed wastes and impurities?

Silence.

FARNSWORTH: Do you see my point now? At some stage in your high school textbook, there was a phrase, maybe it was underlined or boldfaced, that said, "The kidney purifies the blood" or something like that. So you memorized it. Then the teacher asked you a question on an exam—"What is the function of the kidney?"—and you gave back your memorized answer, without really understanding what you were saying, and you got an A on that exam. You don't really know how the kidney works, but you can call back a memorized phrase. Well, in Biology 100, we will not ask "What is the function of the kidney?" We will ask you "Would you expect the kidney of a freshwater animal to be bigger or smaller than the kidney of a saltwater animal of the same size? Explain your answer." A plain memorizer would be murdered by that kind of question, and since I don't want to see anybody murdered, I guess that's why I don't like highlighting. It is okay to memorize vocabulary—as a matter of fact, there's no other way to master it—but I don't want you to try and memorize a description of a concept. I'd rather have you think about it a little, and that's why I encourage active studying. Any other questions here?

STUDENT: Doesn't active studying take a whole lotta time?

FARNSWORTH: Yes. Absolutely. That's where the eight hours a week comes in. It would take you only a couple of hours to read all the material in the conventional way. You have not really studied, if all you've done is read the chapters over. You see, you really do have to make some priority decisions. There are a lot of other subjects you could take that would involve a lot less time, but there is so much material in biology, you really just have to bite the bullet. Next question.

STUDENT: How about studying from old exams?

FARNSWORTH: That one's tough. I assume, naturally, that you're talking about old exams for a course in which the instructor makes them available. Certainly, there's an advantage. You can get an idea of the "style" of an exam, its format, and the level of difficulty of the questions. There are some pitfalls, though. There is a temptation to think that because there are questions on a very specific subarea of a topic, that all you have to do is study that subarea. Very dangerous assumption. Exam questions come out of a rifle, not a shotgun. What the instructor is interested in is your general knowledge of a *topic*, not any one subarea. So if there is a question this semester about a particular subarea, in all probability there *won't* be a question about it next semester. Another pitfall is that you might see a question on the current exam that is very similar to an old one. The problem is that you have a mind-set about the old question and its answer, and you might not notice the crucial difference in the new one. Bottom line is that old exams are a mixed blessing. Now if there are no other questions, let's consider the lecture notes.

There's a lot more variation in approaching lectures than there is in studying texts. A book is a book, no matter what, but every instructor is different, and I want to caution you that the things I say about taking lecture notes in my course may not necessarily apply in another course. I'll try to give you some generalized hints, though, that you can use to figure out what is best in other situations.

There are two problems in a lecture; listening and writing. How much writing you have to do depends on how much the material the lecturer is giving you is duplicated in the text. If most of it is from the text, then your notes can be very simple, just reminders of general topics. If none of it is from the text—for example, if the lecturer is giving you material so new that it isn't *in* any text, or is giving you

controversial material that the book doesn't want to touch—then your notes are going to have to be much more complete because you have no alternative source of information. Since you are bound to run into this situation sooner or later, it will pay you to learn how to take detailed notes. Ideally, your instructor should tell you what the relationship is between lecture and book, but if he or she doesn't, you'll have to do the best you can from the syllabus.

In Bio 100, I give you lectures on text material only when long experience has told me that it is easy for you to get bogged down on a particular subject because the text presents so much factual material that it is hard to see the central idea. So the lecture is a source of explanations, not a source of facts. You need both.

So, your first job is to figure out whether to take down everything, or be more selective. Sometimes, as in Bio 100, the instructor will tell you. Other times, you'll have to figure it out yourself. It will take you a couple of lectures, and some reading of the text, to do this.

If you are going to have to take down everything, some kind of shorthand is helpful. It doesn't have to be a big deal like the system secretaries use. Just leave out the vowels, and you can speed things up. For example, you can figure this out, can't you? (*He puts a transparency on the projector that says, "U cn red ths prty wel, cnt u?"*) Don't try this kind of abbreviation with the technical words, however—too easy to misspell them later.

So, during the lecture, you're going to use this fake shorthand to keep up. Now, here's an important point, very important. At the first possible moment after the lecture, when you have a free 15 minutes or so, go back and clean your notes up. Finish the drawings. Write out the abbreviations in full. Your goal should be that you could give these notes to a friend and the friend could understand them. Now, I know what you're thinking. You're thinking, "Why should I go to that bother when *I* wrote them—I'm the one who's going to read them, and I ought to be able to figure out what I wrote." Ah, but remember, you're going to be reviewing these notes for the final, four months after you've written them. Are you really going to remember what this (*writes "chndrspndylts"*) means, out of context, 120 days after you wrote it? Probably not. I'm not saying you have to rewrite your notes completely, although many instructors recommend it; just edit them so they are perfectly understandable. This is a 15-minute investment that will pay big dividends later. Besides, maybe you can sell your notes to your friends who don't come to lecture. Okay, any questions about lecture? Yes?

STUDENT: Yeah, how do you stay awake?

FARNSWORTH: (*Laughs*) I hope, sir, you are not speaking about *my* lectures. Sure, sometimes that is a problem. You have an eight o'clock class; you've been up 'til three. It's hard. A good lecturer is the biggest help, but that's outside your control. Lecturers have to go to lectures too, to keep up their professional skills, and some of those lectures are as good as a couple of sleeping pills. It sounds silly, but I wiggle my fingers a lot and every couple of minutes take a couple of good, deep breaths to get a little oxygen in. It doesn't help a whole lot, but it does help some. The biggest help is a lot of rest the night before, but that isn't always possible. That's why I don't mind too much if I see an occasional student dropping off. Snoring,

that's a different story, and if the whole class starts to go under, then I know I've got a problem someplace. All right, if there are no more questions about lectures, let's go on to consider examinations.

There are all sorts of exams, but in a big science course like this, a very common kind of exam is the multiple-choice exam, and that's mostly what I'll be talking about. Before you approach taking *any* exam, however, you want to have a feeling for whether you're just going to have to use your memory, or you're going to have to solve problems. If it is going to be *all* memory, it might pay to pull an all-nighter. Fatigue doesn't seem to have that much of an effect on memory. If you are going to be asked to think, or solve problems, it is absolutely imperative that you get enough rest. Fatigue dramatically increases the probability of error in problem solving. In Bio 100, you'll have to solve problems, so that means the night before the exam, go to a movie. Seriously. By that time, it is too late to try to figure out a concept you haven't gotten, and you'll just put yourself into a panic.

Attitude is very important going into an exam. If you tell yourself you're going to flunk it, you probably will. It's a self-fulfilling prophecy. Saying that it's so makes it so. Try not to think of the *outcome* of the exam. Think instead about the questions themselves. Even if you haven't studied as much as you should have, hey, maybe you'll get lucky. Positive attitude is everything. That's true here as well as in sports.

So now you walk into the exam with your positive attitude, and what's the first thing you're going to do when you get the exam? That's right, read the instructions. Count the pages—make sure that you didn't get an exam with the last page missing. Once you've checked these mechanical details, you're ready to start the exam—and I'm going to give you a couple of tricks.

First of all, you have to know if the instructor penalizes you for guessing. Some do, some don't. If there is a penalty, they usually subtract the number of wrong answers from right, or something like that. In Bio 100, there's no penalty, so make sure you at least try every question. Look at question 1. If it is at all difficult, or

will require time to set up, skip it and go on to the next. Go through the whole exam this way, answering only the easy, quick questions. Why do this? Because if you should run out of time, you don't want to leave a whole bunch of easy questions unanswered at the end of the exam. In addition, you will start out on a positive note. After the easy ones are done, go back and try the tough ones, and here are some tips for tough questions.

If you can rule out a couple of answers by the process of elimination, you will be left with three or, even better, two possible answers. If you don't know the correct answer right away, if you have uncertainty, your first guess has a better chance of being correct than your second guess. The reason for this seems to be that the correct answer is probably lodged someplace in your subconscious, and just kind of pops up, like a free association, the first time you read the question. Now I didn't say that your first guess is always correct—it is more like you have a 60 percent chance of being correct, instead of the 50 percent chance you would have if it were a true random guess between two choices. Still, why throw away an advantage, even if it's a small advantage?

Another trick you can use on a multiple-choice exam, which unfortunately won't work for you in Bio 100 because I don't write the exams this way, has to do with certain peculiarities about the way questions are written. If you have five choices, A, B, C, D, and E, and you can't decide between any of the answers, if it's a pure guess, guess C. It turns out that C is the correct answer about 28 percent of the time, instead of the 20 percent you would expect. The reason for this is that instructors are often hesitant about making the correct answer A, because they figure it will be too conspicuous. Same thing for E. A variant on the theme is that the correct answer tends to be the longest one, if the answers are of different length. Don't try to follow these rules in Bio 100, because I don't write the questions like that.

Another important thing to remember—many multiple-choice exams are graded by computer, which often means you have to transfer an answer from a worksheet to an answer sheet. Before you turn the exam in, be sure to check that you didn't make any errors in transcription. Okay, any more questions about exams? Yes?

STUDENT: Any advice on essay exams?

FARNSWORTH: Well, we don't have essays in Bio 100 unless you ask for them, but I'm sure you will have essays in other courses. Preparation for an essay isn't really any different from preparation for a multiple-choice exam, but the actual exam technique is different. There is one piece of information you have to find out about yourself first. You have to know how much you can write in a given period of time. Can you write 100 words a minute? Fifty words? You will have a certain number of essay questions on an exam, and you can figure out how many minutes you can afford to spend on each question. Let's say you have five essay questions for an hour exam. Fifty minutes, really, since the class isn't an hour long. That means that for each question, you have 10 minutes to organize what you're going to say and get it down on paper. If you know in advance that you can physically write four good-sized paragraphs in 10 minutes, you can structure your answer so that it has an introductory paragraph, two paragraphs of information, and a concluding paragraph. It is important to remember that the answer that can't be read can't be graded, so don't be too ambitious about how much you can write. It is also important to remember that an answer should have a beginning, a middle, and an end. It shouldn't just sort of die away without concluding something.

Another thing to remember is that in science courses, BS, garbage that's there just to pad the answer, mostly earns you negative points. The instructor has a lot of papers to read through. The teacher who has to wade through piles of garbage in your answer is an unhappy teacher. Keep your answer limited to stuff you understand.

Okay, I don't know about you, but it's getting late for me, so if there are no more questions, let's call it a night. (*A few students come up to ask private questions—the rest straggle out sleepily.*)

Day 3

Simple Chemical Reactions

Professor Farnsworth enters, wearing a Harris tweed jacket, Explorer's Club tie, and gray flannel slacks. He is quite spiffy.

FARNSWORTH: Good morning. I see some familiar faces from the study skills workshop last night. So how are you today? Comfortable? I hope so because we have a full session, and I want you to be paying close attention. The material I'll talk about in this session is covered in much greater detail in your textbook, but what I want you to do now is feel easy with the ideas. There's nothing here that will be terribly difficult to understand, but we'll still take it one step at a time, nice and easy.

This is a biology course, but a lot of modern biology is really chemistry in disguise. I realize that for some of you that may not be the best news you've heard all day, but I'll surprise you, I think, with how logical it all is. Many of you have had high school chemistry, so you're not going to faint if I mention the word *bond*, but I don't expect you to remember many facts, and I'll refresh your memories when it's appropriate. Okay, let's roll. (*Places blank transparency on overhead projector.*)

Biological chemistry is a special kind of *organic* chemistry. **Organic chemistry** is the chemistry of the carbon atom. There are certain properties that carbon has that make it particularly suited for use in living things. Probably the most important of these properties is its ability to bond to other carbon atoms, thus making long, complex molecules. These big molecules are called **macromolecules**. *Macro* just means "big." For us, the importance of macromolecules is not so much that they are big, but that you can make them in so many different forms. If you have a sodium atom and a chlorine atom, and you try to combine them, you're going to end up with salt no matter what you do, but if you have some carbon, hydrogen, and oxygen atoms, you can make literally trillions of different kinds of compounds.

Nobody knows exactly how many different compounds there are in a living cell, but a reasonable guess might be somewhere between 10,000 and 100,000. With carbon, we can easily make all the different kinds of compounds we might need. Carbon is not the only atom that can form macromolecules. Silicon has similar properties, and it is possible to imagine another world with slightly different conditions that might have evolved a silicon-based life-form. Who knows, maybe living rocks or something. That didn't happen on earth, though; carbon is our

43

fundamental element, so we might as well resign ourselves to learning something about it.

Most of the important macromolecules are called *polymers*. A **polymer** is a compound that is made up of subunits. In most cases, the polymers in biological systems are composed of a relatively small number of different kinds of **monomers**, as these subunits are called. Since I love trains, I'm going to show you how a freight train is like a polymer (*puts transparencies on projector*).

Here we have a freight yard that has all the kinds of cars we need for our trains. On track 1 are the engines. On track 2 are tank cars full of Jack Daniel's whiskey. On track 3 are flatcars loaded with lead bricks. On track 4 are livestock cars full of butterflies. On track 5 are cabooses. One engine can pull only six cars, and according to the union, we can use only one engine per train, so that means that every freight train has six cars. One of these cars has to be the caboose, and as everybody knows, the caboose has to go on the end of the train. We have, then, only three different kinds of freight cars, and each train can then have five freight cars plus a caboose. Now, how many different kinds of freight trains can we make up with the three kinds of cars and five cars total? Well, the exact number is not important, but as you can easily see, it is more than three and more than five. A lot more. We could have five whiskey cars. Three butterfly cars, a whiskey car, and a brick car. Four brick cars and a butterfly car. So, you could make up lots and lots of different kinds of trains with only three different kinds of cars. Let's call the different kinds of cars *monomers*, and the train a *polymer*.

Let me ask you this. Would the different kinds of trains you could make up have different *properties?* Which would be easier to pull up the Rockies, five brick cars or four butterfly cars and a Jack Daniel's car? The butterfly–Jack Daniel's combo,

naturally. If you were going around a tight curve, could you go around faster with five butterfly cars or five Jack Daniel's cars? The butterflies, of course—the Daniel's cars might slosh themselves right off the track. Every train, then, will have different properties depending on what kind of monomers, er, *cars* make it up.

Now, here's a question that has a less obvious answer. Do the properties of the train change *depending on the order of the cars behind the engine?* Let's say we have two trains, each with two brick cars and three butterfly cars. One train has both the brick cars right behind the engine; the other has the brick cars in front of the caboose, with the three butterfly cars between the brick cars and the engine. Both trains go up and over a steep hill. Remember now, both trains have exactly the same *kinds* of monomers, I mean, cars, but they're arranged differently. The first train, with the bricks right behind the engine, crests the hill with no problem. The second train, with that huge weight at the *end* of the train, gets halfway over the top of the hill and breaks in half because the weight at the back of the train is pulling in a

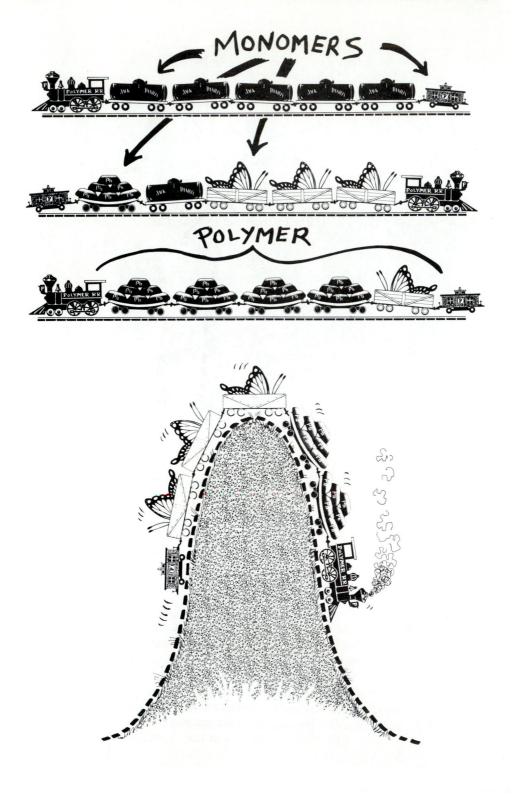

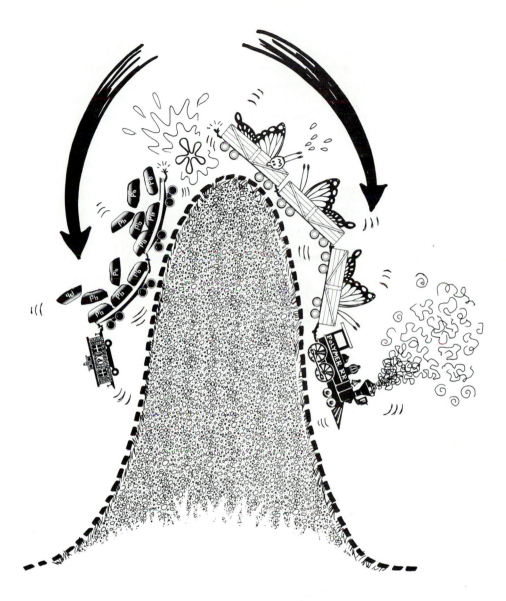

direction opposite to the pull of the engine at the front of the train, and the couplers between the cars in the middle fail. Similarly, the two trains would behave differently going around a curve—there would be more of a crack-the-whip effect with the weight in the back.

Suppose, though, we weren't talking about a train made up of brick cars and butterfly cars, but a train made of brick cars and whiskey cars—would the order of the cars be as important? No—the brick cars and the whiskey cars have similar weights. Therefore the importance of order is at least partially dependent on the properties of the monomers, er, cars.

So we've seen that the characteristics of a polymer depend not only on the *kind* of subunits, but on the *sequence*, or *order*, in which they are arranged. And as we will see later, it is true of many kinds of molecules, not just polymers, that the arrangement of the atoms, as well as the types of atoms in the molecule, is important.

All right, now we're ready to look at real macromolecules. The first class of molecules is called **carbohydrates**. *Carbo-hydrates*—what does the word mean? *Carbo* means "carbon," and *hydro* means "water." Carbon and water. In a chemical formula, water is H_2O, so a carbohydrate would have the general formula CH_2O. I said "general formula"—what does that mean? Well, one kind of carbohydrate might actually have that formula—one carbon, two hydrogens, one oxygen. Another kind might have two carbons—so how many hydrogens would it have? It would have four, so its formula would be $C_2H_4O_2$. If we had a carbohydrate that had *six* carbons, what would its formula be? You got it—$C_6H_{12}O_6$. By a strange coincidence, the first carbohydrate we will look at in detail has six carbons, and it is called *glucose*. Before we look at glucose, there are a couple of other things I have to tell you about. Here is a formula—$C_6H_{12}O_6$. Is there only one possible compound that could have that formula? No! As we saw with the freight train, it is not only *what* you are made of that is important, but also how your parts are arranged. In organic chemistry, carbon chemistry, the way we show you how the atoms are arranged in a molecule is with something called a **structural formula**. This doesn't show you the whole structure, because a drawing is in two dimensions, length and width, and the real molecule has three dimensions, length, width, and depth, but at least it shows you a lot more than the chemical formula.

If you look at the structural formula for this six-carbon carbohydrate, glucose, you can see that a chain of five carbons is connected by an oxygen to form a ring and the sixth carbon is off to one side. This sixth carbon is interesting because it is part of what is called a **functional group**. That OH hanging down there off the

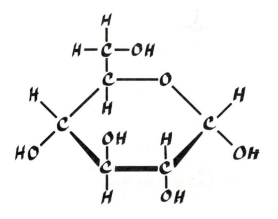

Structural formula
for Glucose

bottom is a functional group, too. Functional groups are attached to a basic molecule, in this case, a five-carbon ring. The nature of the functional groups and where they attach to the main part of the molecule determine the properties of the whole molecule.

It takes a long time to draw all those C's and H's, so there has developed a kind of shorthand for structural formulas that I will use from now on. Whenever there is just an H attached to the carbon, we show it as a short line. Where there is a carbon in the main part of a ring, we show the location of the carbon by the intersection of four lines, representing the four bonds that carbon can have. In this next picture you can see the glucose molecule written in this shorthand. Now over on the right, there is a drawing of another six-carbon sugar, galactose. Both of these formulas have the same chemical formula, $C_6H_{12}O_6$, but as you look closely at the drawings, you can see that the atoms aren't arranged the same way. This different arrangement gives the molecules different properties.

Let's start now to make a macromolecule. Let's say I had *two* of these glucoses and I wanted to combine them to form a larger molecule that would be essentially two glucoses stuck together. First of all, what would I call such a thing? What kind of compound is glucose? A carbohydrate, correct, but is there a common name? Sure, *sugar*. The technical name for a sugar is a *saccharide*. Glucose is a single molecule, so it is called a *monosaccharide*. If I fastened two monosaccharides together, what would I have? A *disaccharide*. Suppose I fastened a whole SHIP-LOAD of monosaccharides together, what would I have? All together now, a *POLYSACCHARIDE*.

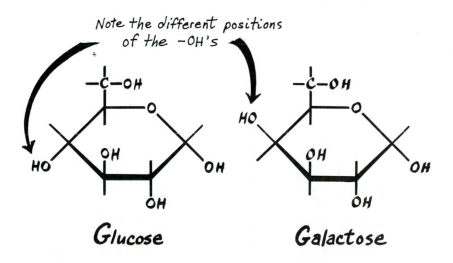

Glucose

Galactose

Okay, I'm not greedy; all I want is a disaccharide for now. How am I going to make it? My next figure shows two glucoses side by side. Look at them for a minute. How am I going to get them together? I've got a functional group here—the OH functional group. The OH group is called a *hydroxyl group*, and look here, two of the hydroxyl groups (one from each glucose) are close to each other. Suppose I

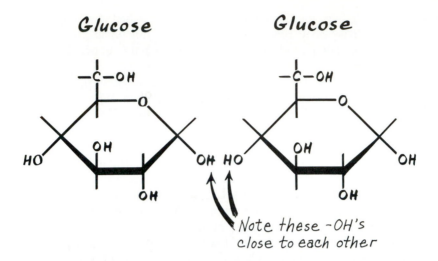

Note these -OH's
close to each other

were to take a little, tiny hatchet and *split* one of these hydroxyls clean off its carbon, and SPLIT the second hydroxyl right in two. What would I have? Floating around here, I would have a hydrogen and a hydroxyl. Suppose they sucked right up to each other and bonded together? I'd have a water molecule, good ol' H_2O. Now, let's go back to the glucoses. On the left-hand one, we have an oxygen with a vacant bond, and on the right-hand one, we have a carbon that has a vacant bond. If we pushed these two glucoses together, the carbon would latch onto the oxygen, and we would have a new, larger, stable molecule that was made out of two monosaccharides. In other words, a disaccharide. This particular disaccharide, made from two glucoses, is called **maltose**; it got its name because it is found in sprouting grain, which is called *malt,* which is used to make beer, which is perhaps why I know more about maltose than other disaccharides.

It is time to attach some names to this process. Because we are building a larger molecule out of smaller ones, it is called a **synthesis**—*syn* meaning "together." It is a particular kind of synthesis. Because we removed the constituents of a water molecule and made a water molecule from them, the process is called a **dehydration synthesis**.

Suppose, now, we wanted to go the other way. Suppose we had a maltose and for some reason wanted a couple of glucoses. We could create them fairly easily. We would first need to split the maltose molecule, then replace an H^+ on the vacant O, and replace an OH^- on the vacant C. Where could we get these H^+ and OH^- groups? By splitting a water molecule and using the by-products of the splitting, an H^+ and an OH^-, to reconstitute two glucose molecules. Because we are basically making smaller molecules out of bigger ones, we call this a **digestion reaction**. Digestion in your guts is just making lots of small ones out of a few big ones. Because we used a water molecule as a source of ions—remember *ions* are charged particles—and we split the water molecule to get them, this is called a **hydrolysis reaction**—*lysis* meaning "to split." This is then a hydrolytic digestion. Okay, any questions? Yes, blue T-shirt on the side?

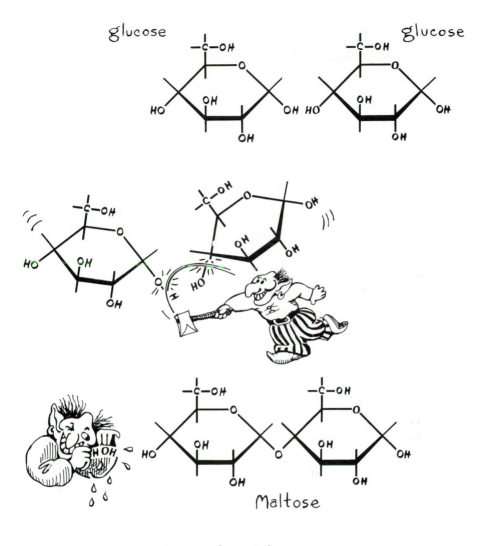

Dehydration
Synthesis

BLUE T-SHIRT: How do you split the water?

FARNSWORTH: I was afraid you were going to ask me that because I can give you only a partial answer now; for a complete answer, you're going to have to learn a little more. Still, let's see how far we can go for now. There are bonds that hold the atoms of the water molecule together, right? So when I say "split the water molecule," what I really mean is "break the bonds." How do you do that? Well, this

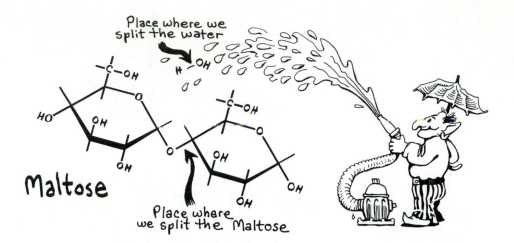

Place where we split the water

Place where we split the Maltose

Maltose

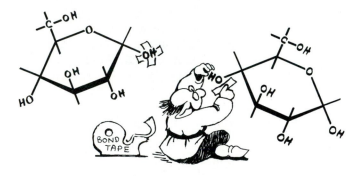

Glucose

Glucose

is not exactly how it works, but this example will be close enough for right now. Let's say you have two bar magnets and you kind of push them around at random on a table with your fingers. Sooner or later, the opposite poles will come close enough to attract each other, and—*sproinnggg!*—they'll pull each other together and stick. Now we have these two little magnets stuck, or bonded, together, and if we push them around on the table, they'll go together. So far, so good. We have these two magnets stuck together, but now we want to separate them. If we pushed them apart just a little bit, what would happen? They'd instantly pull back together. But if we pushed them apart a little farther, the magnetic force wouldn't be strong enough to pull them together and we could pick them up separately and do what we wanted with them. Now, would the only way to break them apart be to grab each one by the end and pull? No, maybe we could just shake the table they're on, and after a while they would vibrate apart. The harder we shook the table, the greater the chance they would shake apart, right? So we could either shake them apart or pull them apart. Okay, now we're ready to talk about bonds and water. For reasons that we'll talk about later, molecules shake, or vibrate, all the time. If we increase the amount they vibrate, we can increase the probability that they will shake apart. One way to do that is by heating the molecule. Another way we could split the molecule is somehow to reduce the amount of energy needed to pull the atoms apart. If we did that, we could shake the molecule just a little bit and it would

break apart. There is a process that can do just that—it's called **catalysis**, and we'll speak of it again, but right now let me see if I can put this together.

Here you have, then, a water molecule. It is vibrating at a certain rate, which at body temperature is rarely enough to cause it to shake apart. We could make it more likely to break apart by itself by heating it, but the amount of heat required to do that is way beyond anything a living organism could stand. So we have to take the opposite approach—changing the amount of energy required to push the atoms apart. This we do by adding a *catalyst*, which will have an effect only on a particular kind of bond. So to split our water molecule into an H^+ and OH^-, we add a catalyst that works only at that particular spot on a molecule, and it then conveniently proceeds to shake itself apart for us. We would then want to grab the H^+ and OH^- before they could re-form into water, but there are ways of doing that.

So to wrap up your question, the water molecule shakes all the time, but not hard enough at room temperature to shake itself apart. If we don't want to raise the temperature of the molecule to split it, we can add a catalyst which will reduce the amount of energy required to break the bonds holding the atoms together. The molecule then *will* proceed to shake itself apart at room temperature, breaking at the particular point where the catalyst acted. You would use a different catalyst to allow the original maltose to split apart at moderate temperatures. Does this more or less take care of your question? Good. We'll be coming back to this later, but this will get you started. That was a very good question.

Now that we understand everything there is to know about carbohydrates, let's look at the next class of biologically important molecules, *fats*. What do you use fats for, in the body? Anybody?

STUDENT: Energy.

FARNSWORTH: Good, but that actually isn't the major use—you can metabolize carbohydrates for the same purpose. The major use is structural. Fats are raw materials for a lot of hormones, and they're also extremely important parts of the cell membrane. So you wouldn't want to get rid of *all* your ugly fat, because if you did, all your cells would disintegrate. With that in mind, let's take a look at a typical fat. (*Puts transparency on projector; see page 56.*)

A fat is made of a kind of alcohol called *glycerol*, sometimes called *glycerine*, bonded to three *fatty acids*, as they're known. There are lots of different kinds of fatty acids, and the different kinds of fats are determined by which fatty acids make them up. A fatty acid is a long-chain hydrocarbon that has a *carboxyl*, or COOH functional group, at one end.

So, let's say we have a glycerol molecule, and we have three identical fatty acids next to it—it doesn't matter which ones—and we want to glue them together into a single fat molecule, whose fancy name is *triglyceride*. How do we do it? Well, this is a synthetic reaction, right? We're building up a molecule, and lo and behold— does this sound familiar? OH groups next to each other. We saw this when we were making disaccharides, except there we had only one pair of hydroxyls next to each other. Could we bond these fatty acids to the glycerol the same way, by a dehydration synthesis? Damn right, we could, and that's just what we'll do. We'll pull three water molecules out, and presto! A triglyceride. The triglycerides, along with oils and waxes, are collectively called *lipids*.

Now, suppose you were very angry with me after an exam and wanted to blow up my car. You could do it if you had some nitroglycerin, but to make the nitroglycerin you would need glycerol. You don't have any glycerol, but you do have a nice source of camel fat. The camel fat is a triglyceride, so how could you break it apart into fatty acids and glycerol? Same way we broke the disaccharide down into monosaccharides, by performing a hydrolytic digestion, except you would use three waters instead of one. If you think I'm going to tell you *exactly* how to do this, you're crazier than I am—my car isn't even paid for yet, and a lot of you are going to be angry with me after the exam!

So now we have two classes of compounds, carbohydrates and fats, that can be assembled and disassembled in the same way. We are now going to look at a third important group of biological chemicals, the *proteins*, and see what we can do with them.

All the nutritionists say you need protein. What for? Anybody?

STUDENT: Energy.

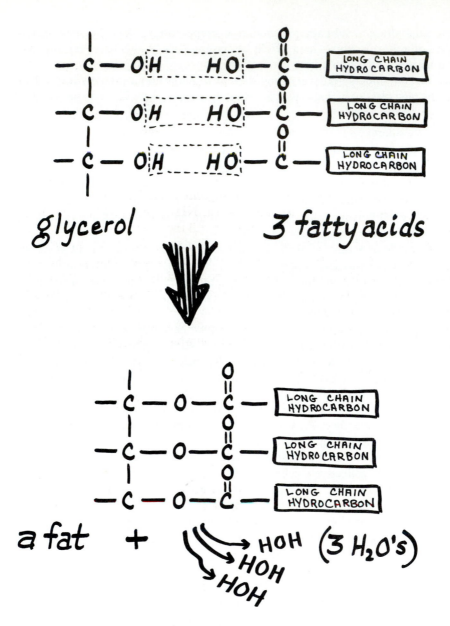

glycerol

3 fatty acids

a fat + → HOH (3 H₂0's) HOH HOH

FARNSWORTH: Okay, you *can* use protein as an energy source, but only as a last resort in starvation. You use carbohydrates and fats for normal, day-to-day energy requirements; proteins are used for something else. They can be used for body structure—muscle is mostly made out of protein—but the most important use for protein is as biological *catalysts*. Remember catalysts? The things that reduced the amount of energy needed to break bonds, and thus speeded up the rate of a reaction? Well biological catalysts are usually made out of protein and are called **enzymes**.

enzyme is a biological catalyst used to make a reaction go at normal temperatures. Just about every different reaction in a cell requires a different enzyme, and we've already seen that there are tens of thousands of different reactions in a cell, so that means we need tens of thousands of different enzymes. The thing about proteins that makes them so good as enzymes is that it is very easy to make up lots of different kinds because they are macromolecular polymers, like the polysaccharides, but in this case, the monomer is something called an **amino acid**.

There are 20 different kinds of amino acids used to make proteins. If you look at a drawing of the basic structure of an amino acid, at one end you will see a familiar group, a COOH. You remember what that is? That's right, a carboxyl. The carboxyl here makes the molecule an **organic acid**. Then over on one side there's this group with an N and two H's. An NH$_2$ group. That is called an **amine** group, which makes this an amino acid. All the amino acids have this basic structure. But up at the top there's this group that says "R." R means "radical," which in this case means any functional group. There are 20 different functional groups we could put up here. This R could be very simple. For example it could just be an H, in which case we would call the amino acid *glycine*. If it were a CH$_3$, we would call it *alanine*. Or we could have a big complicated side group with maybe three or four more carbons and a couple of nitrogens. Each of the amino acids will have different properties, depending on which side group it has. Some of

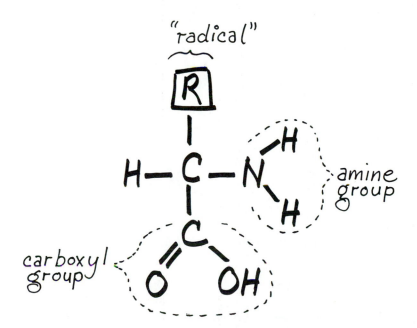

Amino Acid

the side groups have ring structures, as the glucose had, and are called *aromatic* amino acids. They're called **aromatics,** not because they necessarily stink, but because in the early days of chemistry, when chemists were first studying ring structures, a lot of the first ones they examined *did* have strong smells.

So, now, our basic monomer for a protein is this amino acid. To make the protein, we want to fasten these amino acids together. If we put two amino acids belly

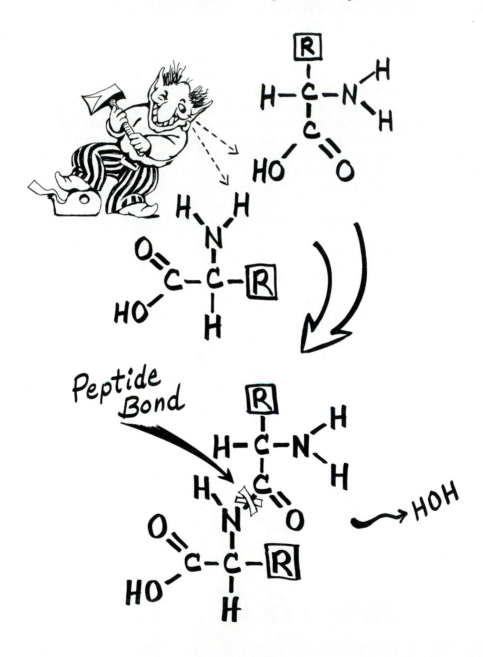

to belly, by golly, we've got an H and an OH right next to each other, and is there some bright person in the class who could tell me what to do with them?

BRIGHT PERSON: Dehydration synthesis!

FARNSWORTH: Exactly. But when we pull out the water, you see that we're not attaching an O to a carbon as we did with the fats and polysaccharides; instead we have a C next to an N. A C bonded to an N forms what is called a *peptide bond*. So, two amino acids joined by this peptide bond are called a *dipeptide*. A bunch of amino acids linked together is called a *polypeptide chain*. One or more polypeptide chains arranged in a specific way are called *proteins*.

There is one other major group of biological molecules called the *nucleic acids*, but I'm going to postpone talking about them until we talk about how cells transmit information to each other.

Let me see if I can wrap this up for you now. We've seen that many of the important biological molecules are polymers: large molecules made up of repeating smaller subunits called monomers. Large molecules are made from smaller ones by synthetic processes; small ones are made from big ones by digestion. The removal or addition of a water molecule is often important in synthesis and digestion. The properties of a macromolecule are a function of both the kind of monomers of which it is made and the order in which the monomers are arranged.

Now, before we finish, and by way of a little preview of next time, I'll ask you to think back for a second to some of the formulas and equations I drew. Was there something I left out? I mean, the equations were balanced, right? What very important factor didn't I consider? Yes, pink blouse on the left—by the way I hope you'll forgive me for referring to you all by your articles of clothing rather than by name—it's too early in the semester for me to know many of your names. So now, what did I forget?

STUDENT: Energy. You didn't say whether the reaction gives off or requires energy.

FARNSWORTH: Exactly! In a reaction, you are dealing not only with mass but with energy. Next time, we will see what role energy plays in biological chemistry. Have a nice weekend.

Day 4

Glycolysis and Fermentation

Professor Farnsworth enters. He is wearing a navy blazer, dark-gray slacks, and cordovan loafers. His tie looks like a standard yacht club model, dark-blue with little flags, unless you look at it closely, in which case you would see that the little flags are actually portraits of Goofy.

FARNSWORTH: Good morning. A nice weekend, yes? Burn off a lot of energy? Good, because that's what we're going to talk about today. Energy. How you get it, how you use it, and how it is used in biological systems. Before we do anything, however, we ought to decide what energy is. Anybody?

STUDENT: Power.

FARNSWORTH: Anything else?

ANOTHER STUDENT: Work.

FARNSWORTH: All right, I was afraid of that. You're using the terms interchangeably. They're not the same thing at all. First of all, energy. **Energy** is the capacity to do work. Okay, what's work? **Work** is the movement of an object against some force. The object can be a physical thing like a brick, or it could be something subatomic, like an electron. So if you push a car, you're using energy to perform work. You're moving the car, against the resistance of friction. **Power** is the rate at which work is done or energy is expended. There is always a time component with power. How far can you push the car in a minute, for example.

Energy can take different forms, and you can change energy from one form to another, but you cannot either make or destroy energy—you can only change its form. But here's a complication. Whenever you change energy from one form to another or you do work with that energy, some of that energy may be transformed into a form which cannot be used for more work. Take a car, for example. You put gas in the engine, the gas is burned in the cylinders, and some of the chemical energy in the gas is transformed into work—the car moves. But is all the energy in the gas transformed into movement? No, some of it is converted into heat. Now, you could use some of that heat to warm up the inside of the car, but what happens to the heat then? It gets transferred to the air surrounding the car and heats it a little bit. What happens to the

heat in this air? Well, it might get moved around a little, but eventually it will get lost to outer space. Where does it go from there? The universe absorbs it. What happens to it after that is a philosophical question; you would have to ask a priest or a physicist what happens to the energy after that. From a biologist's standpoint, all we have to know is that *every* time energy is transformed from one form to another or used to perform work, some of that energy is lost as heat.

Now we are ready to make some progress here. Last time, somebody mentioned that we hadn't considered energy in chemical reactions. Let's take a look at that.

Let's say we have a chemical reaction that releases energy. This reaction might involve the formation, and breaking, of bonds. Three things to keep in mind. One, energy can be stored in bonds. Two, it takes energy to break a bond. Three, energy is released when a bond is formed. How, exactly, might this work?

Start with a chemical reaction that involves the breaking of a bond. In many biological processes, the breaking of that bond will eventually release energy that we can use for some useful purpose. However, as we will see, sometimes you have to feed a little energy into the system to get the bond to break. How, exactly, does this work? (*He reaches behind the podium and pulls out a couple of basketball-sized spheres.*)

I have here a couple of bowling balls, but they are very unusual because they have only two holes instead of three. If we drew a cross section of the balls, we would see that the holes are angled to each other. Now, for reasons of my own, I want to fasten, or bond, these two balls together. If I grab a couple of steel rods (*he reaches behind podium again*), I could stick a rod in each of the holes in one of the balls. Then I could stick one of these two rods in one of the holes in the other ball. Finally, I could take the other rod, stick it in the second hole of the other ball, and—oh, hell, it won't fit in the last hole, because the holes are angled away from each other. I'll have to think of something else.

Look what I just found! A couple of really stiff coil springs, and see here, they just fit in the holes. I'll stick 'em in both holes of one ball, and then if I (*grunts and strains*) squeeze really hard, I should be able to bend the springs together enough to get them in the holes of the other ball. There! Done. Have I fastened the balls together? Yes. Important point now. Did I release any energy to bond them? Also yes. Where was the energy released? By my muscles, when I squeezed the springs together. That's why you get hot when you exercise. Now, was *all* the energy I used to form the bond released? No. Some of it is stored in the bent springs. Now, look at those springs. I have put energy into them, but where is that energy now? Are the balls doing anything? Jumping around? Glowing in the dark? No. The energy is stored in the bent springs, waiting to be released. This is very similar to what happens in a chemical bond. You release energy to get the bond to form, but some of the energy sits quietly there in the bond until you release it somehow.

Suppose I wanted to do something with this spring energy now. Suppose I wanted to kill a cockroach with it. Oh, I could pretty much do anything I wanted—produce heat or fire, or make another kind of bond—but as it happens, there is this cockroach walking around the stage that has been bothering me, and I want to trash it. So, I have this little dog bowl here, and I'll put it on the stage and place the lower ball in the bowl, with the holes facing up. You can see that the top ball is then above

the lower one, with the spring bonds in between. The cockroach—you can't see it, but trust me, it's here—is over here about a foot from the bowl. What would happen if I cut the spring on the side opposite the cockroach? You got it! The top ball would whip over and smoosh the cockroach. But how am I going to cut the spring? Can I just do it with my fingers? No, but I have a couple of choices.

I could use a blowtorch to melt one of the springs. Or I could use a big bolt cutter, put some muscle in it, and snip one of the springs. In both cases, I have to expend some energy to break the spring.

Same thing with a bond. To break a bond, you have to feed in some energy, called the **activation energy**, to get things going. When the bond is broken, some of the energy is stored when you form new bonds among the reactants and some is released as the result of the formation of the new bonds that produce the products of the reaction.

The fact that you need activation energy to start many reactions accounts for the fact that a lot of very powerful reactions just don't get started under ordinary conditions. For example, if I poured a gallon of gasoline over the stage here, what would happen? There's oxygen in the air; there's gas on the ground. Why doesn't it go up? Because the reaction needs activation energy, which might be provided if someone should accidentally drop a match into the pool of gas. Fortunately for us, most biological reactions work the same way—activation energy is required to start things. If that weren't true, we might all spontaneously ignite as we sat here. Not a pleasant thought.

Okay, we've had a taste of *energetics*. Let's do something with it. Suppose you have a coal-fired power plant that produces electricity on the south side of town and you want to get that power to the north end. What would you do? Ideas, please?

STUDENT: You could string wires from the plant to the place where you wanted the energy.

FARNSWORTH: Excellent! More ideas, please. (*Long pause*)

ANOTHER STUDENT: Well, this sounds crazy, but you could get a big truck and put batteries on it. You would charge the batteries at the plant, then take them across town and hook them up where the energy is needed. After the batteries were discharged, you'd take them back to the plant for recharging.

FARNSWORTH: Exactly! You must have ESP. But there are two kinds of batteries, aren't there—the kind you throw away, like ordinary flashlight batteries, and the kind you recharge. In this crazy system, what kind would you want to use, and why?

THE STUDENT: The rechargeable kind, because it would require energy to make the batteries, and when you threw the batteries away, you'd waste that energy.

FARNSWORTH: Good guess, and that leads me into what this discussion is all about. As we've seen, a cell requires energy to assemble and disassemble the molecules it needs. This energy comes into the cell in the form of bond energy holding food molecules together. The cell extracts the energy from these bonds by a process we'll talk about in a couple of minutes, but the place where energy is extracted, a cellular structure called a **mitochondrion**, is not necessarily where the energy is ultimately needed. To move energy around inside the cell, we use a kind of molecular re-chargeable battery called **adenosine triphosphate**, or **ATP**.

This relatively small molecule, maybe roughly four times as big as a glucose molecule, has three main parts—let me show you a picture. (*He places a transparency on the overhead projector.*) First, a single-ring sugar called **ribose**; second, a double-ring structure called **adenine**, which is an organic base—we'll talk more about them later—and the most important part, three phosphate groups tacked on to the ribose. Now, the key thing about this little molecule is right there between the phosphates: the bonds. There are two of them. These bonds are called *high-energy bonds*, which is unfortunately one of these confusing terms. *High-energy* doesn't mean that the phosphates are very strongly bonded together. As a matter of fact, they are weakly held, but when the bond is broken, a lot of energy is released.

The ATP is like a storage battery. If we start with an ATP molecule, and we want to get some of the energy out of it, we break the outermost high-energy bond, which gives us a phosphate group and a leftover molecule which we now call **ADP**, or **adenosine diphosphate**. Most of the energy stays with the broken-off phosphate, which we can then bond to the molecule which needs the energy. This process is called **phosphorylation**.

The ADP molecule can then drift back to the mitochondrion and pick up an-other phosphate, which is then bonded to the ADP using the energy acquired from another food molecule. Now you have a recharged ATP, which is released into the cell and picked up by any molecule which needs it.

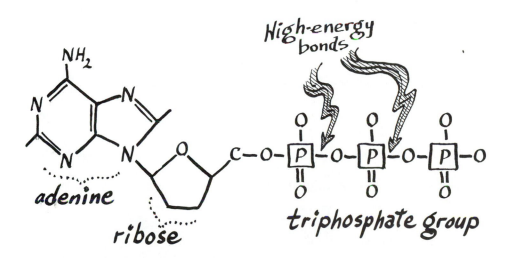

What you have here then is a cycle (*puts transparency on the projector*), in which energy comes into the body in the form of large molecules. In the digestive system, these large molecules are broken down into smaller ones which can get into the cells. Once inside the cell, the small food molecules are almost completely broken down. The energy released in this process is used to fasten a phosphate to an ADP, making an ATP. This can then be discharged where it's needed and the resultant ADP charged up again. Okay, time to stop for questions. Yes, sir, in the Amazing Fishcake shirt on the left?

AMAZING FISHCAKE SHIRT: Yeah, how does the ATP molecule know where to go where it's needed?

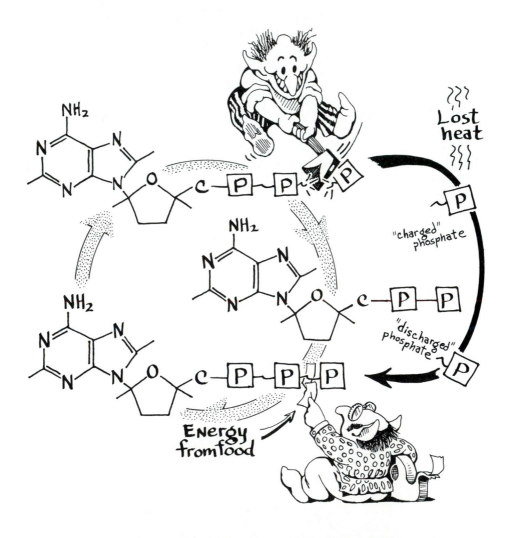

ATP-ADP CYCLE

FARNSWORTH: Ah, good question. It doesn't have to know. There are so many tens of millions of them in each cell that they just diffuse out from the mitochondria all through the cell. There's normally always an ATP whizzing by an ongoing re-action that needs one. There are so many of them that almost all the molecules in the cell are constantly bombarded by ATP's. It's not as though there are a bunch of little railroad tracks going to specific places in the cell that need energy. Okay, another question. Yes, redhead with shades in the back.

REDHEAD: The whole cycle business seems too complicated. Why not just make the ATP's directly, break 'em down totally, and start from scratch with new ones?

FARNSWORTH: Simple reason. You need too many ATP's. In each cell, you use up about 10 million ATP's a second. If you didn't recycle, you would have to man-ufacture about 120 pounds, about 50 kilograms, of ATP a day. More questions? No? Well, now I'm going to need your undivided attention for the rest of the hour, and you have to be awake. Everybody put down your notes, stand up, and stretch. Shake 'em out. (*With evident relief the students follow instructions.*)

Okay, let's go. I told you that the energy to bond a phosphate to an ADP came from the breakdown of small food molecules which entered the cell after preliminary break-down in the digestive system. There is also another way of getting energy for the same purpose, and that is what plants do. They get it from sunlight. For the balance of this hour, and for the next two hours of lecture, we are going to look, in detail, at how animals and plants capture and use the energy contained in chemical bonds and sun-light to produce ATP from ADP. And do you know what that means? That means for the next, let's see, 143 minutes we are going to talk about nothing but CHEMISTRY. What do you think about that? (*Chorus of boos, moans, catcalls*)

I knew you'd love it. It won't be that bad. We'll do it one step at a time, and I'll use a local anesthetic. We're going to start with an example, the breakdown of a carbohydrate to transfer the energy from the carbohydrate to ATP's. Fat and pro-tein breakdown are similar, but a little more complicated. We'll cover them later.

Start with a cube of sugar. Table sugar is a disaccharide called *sucrose*, which is made of one molecule of glucose, a monosaccharide, and one of *fructose*, another monosaccharide. In the digestive system, the sucrose is broken down into the glucose and fructose. Both get absorbed into the bloodstream and then through the cell mem-brane into a cell, any cell. The fructose also gets broken down, but we're not interested in it for now. We are just going to follow this single glucose molecule, okay?

A couple of points to keep in mind. First, this is happening not just in a special organ, or special cells in a particular organ, but in just about every cell in your body. Right at this very instant, for example (*he pinches up a fold of skin on the back of his hand*), in your skin cells, in your bone cells, in your heart, your lungs, everywhere, all the reactions I'm talking about are happening millions of times a second.

The second thing is that I'm giving you a very, very simplified version of what happens. Just the high points, to give you the idea of what is going on. In real life, for every step I talk about, there are 10, maybe 20, intermediate steps. The whole process of breaking down glucose in a cell to extract the energy takes hundreds of steps. Now, why do you think it takes so many steps—I mean, what is the biological advantage of doing it in so many steps? After all, you don't need all those steps—

you can break the sucrose down all in one step. Watch. (*He reaches in his pocket and pulls something out.*) Here's a sugar cube. Pure sucrose. Now I'll just strike a match here, touch the flame to the corner of the cube, and presto! (*The cube begins to darken and bubble.*) Our sucrose is now carbon dioxide, water, and a little ash, and it, ow! (*he touches the pile of ashes*), gave off a good deal of heat. By the way, when you try this, it won't work. You have to dip a corner of the cube into cigarette ashes first. Think about why that's necessary.

Anyway, you can extract energy from sucrose without going through all the rigmarole of enzymes and a million steps. Why doesn't the body do this? Anybody? Yes, on the aisle.

STUDENT: Well, first, you needed a flame to get it started. I guess that means the, what did you call it, the activation energy is high. Where are you going to get a flame inside a cell? And second, all the energy was released in a hurry, so a lot of heat was given off in a short time. If you tried to do that in a cell, the cell would just burn up.

FARNSWORTH: Perfect! By having many steps in which only a little energy is lost as heat during each step, you avoid a dangerous heat buildup. You also have more control of the process. Good, now we're ready to look at the actual breakdown of glucose—let me remind you again that the sucrose was broken down into glucose and fructose outside the cell. (*He puts a transparency on the projector.*)

The first step in the process is called **glycolysis**. *Lysis* means to "break apart." Remember from hydrolysis? We start with a single glucose molecule. How many carbons does it have? Six, right. The first part of glycolysis is *mobilization*, during which we're essentially preparing and reshaping the glucose molecule for other things. First, we take an ATP molecule, strip the phosphate off, and fasten the phosphate

GLYCOLYSIS

Step One: Mobilization

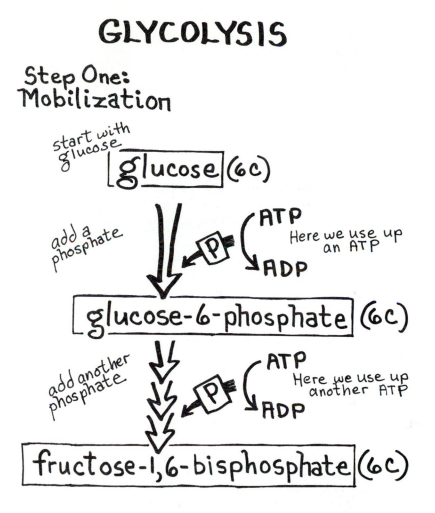

start with glucose

glucose (6c)

add a phosphate

P ← ATP / ADP

Here we use up an ATP

glucose-6-phosphate (6c)

add another phosphate

P ← ATP / ADP

Here we use up another ATP

fructose-1,6-bisphosphate (6c)

to the glucose, making it now *glucose-6-phosphate*. Notice that it has already *cost* us some energy to get things rolling. You have to invest money to make money. Same thing here, with energy. Now we take this glucose-6-phosphate, change its internal structure in a couple of steps I haven't shown, and add another phosphate from another ATP, giving us *fructose-1,6-bisphosphate*. So, at the end of mobilization, we are operating at a deficit of 2 ATP's.

The next stage of glycolysis is called *cleavage*. We split the fructose-1,6-bisphosphate into two 3-carbon molecules, each called *glyceraldehyde-3-phosphate*. Now, this is *very important*. We had one 6-carbon molecule. Now we have two 3-carbon molecules. We are going to follow the fate of only one of these, to avoid repetition, but you have to realize that exactly the same thing is happening to the other 3-carbon molecule—we just won't talk about it, okay?

Our next stage is *oxidation*. Remember, in chemistry, oxidation is not just adding oxygen. *Oxidation* is the loss of an electron, usually along with a proton, from one atom, ion, or molecule (as in this case), to another.

Step Two: Cleavage

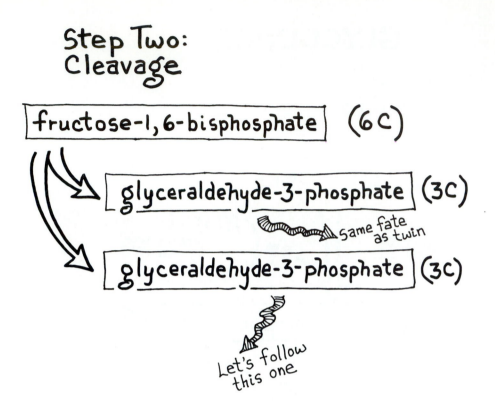

fructose-1,6-bisphosphate (6C)

glyceraldehyde-3-phosphate (3C)

same fate as twin

glyceraldehyde-3-phosphate (3C)

Let's follow this one

What we do here is strip a proton and two electrons off the glyceraldehyde-3-phosphate, and fasten them to an ion called **NAD$^+$**—that's pronounced *N-A-D-plus*. The NAD$^+$, after it picks up the electrons and proton, is said to be *reduced* and is now something called **NADH**. The glyceraldehyde-3-phosphate has lost the proton, so it has been oxidized, and the NAD$^+$ has been reduced. The molecule left after the oxidation of glyceraldehyde-3-phosphate is called *1,3-bisphosphoglycerate*.

This NAD$^+$ is very important. Without it, we'd be stuck right here with all this glyceraldehyde-3-phosphate, and no place to go with it. We need a good supply of NAD$^+$, and in a little while we'll see where it comes from.

In the last stage of glycolysis, we take the 1,3-bisphosphoglycerate and carry it through a series of separate reactions in which we rearrange its internal structure and convert it into something called *pyruvate*, or *pyruvic acid*. In the process, we liberate enough energy to make 2 ATP's.

Now, here's the part where everybody gets confused. We started with *one* 6-carbon glucose. When we went through mobilization, we spent 2 ATP's, so we were *2 ATP's* in the hole. We then split the 6-carbon glucose into *2* glyceraldehyde-3-phosphates, each with 3 carbons. No energy cost to do the splitting. We then oxidized *each* glyceraldehyde-3-phosphate, so that we had *two* 3-carbon 1,3-bisphosphoglycerates. We still have accounted for all 6 of the original carbons in the glucose. At this point, we still have a 2-ATP deficit. We then rearrange *each* 1,3-bisphosphoglycerate to convert it into a 3-carbon pyruvate. *Each* conversion yields

Step Three: Oxidation

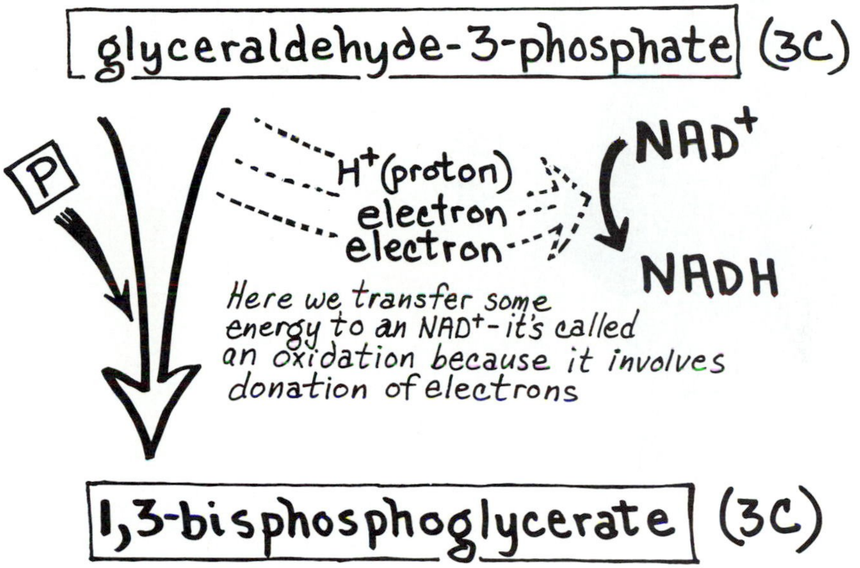

glyceraldehyde-3-phosphate (3C)

H+ (proton)
electron
electron

NAD+

NADH

Here we transfer some energy to an NAD+ - it's called an oxidation because it involves donation of electrons

1,3-bisphosphoglycerate (3C)

Step Four: Conversion to pyruvate

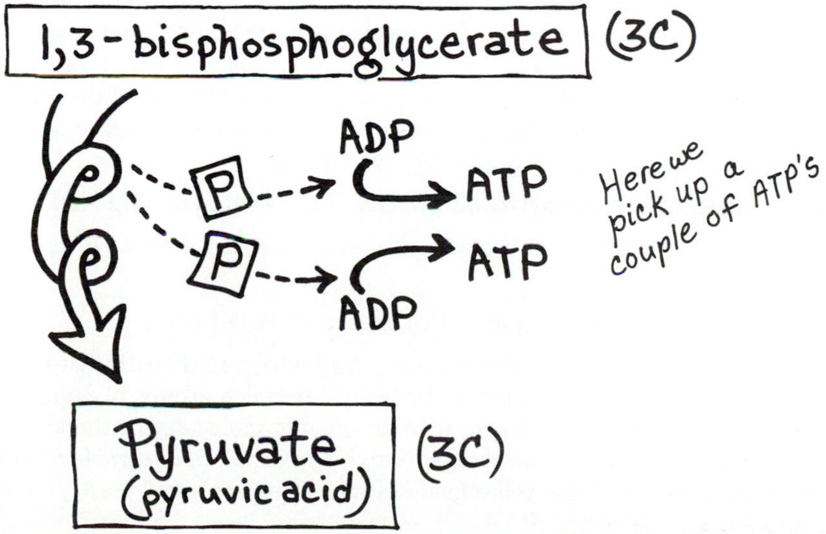

1,3 - bisphosphoglycerate (3C)

ADP → ATP

ADP → ATP

Here we pick up a couple of ATP's

Pyruvate (pyruvic acid) (3C)

us 2 ATP's, or a total of 4. Since we spent 2 ATP's and made 4, we now have, at the end of glycolysis, two 3-carbon pyruvate molecules, and a net gain of 2 ATP's.

Two ATP's is not a very big return on the investment of all that chemical effort. Most of the energy (over 95 percent in fact) that was in the original glucose molecule is still tied up in the bonds holding the pyruvate molecules together. If we could rip up the pyruvate, we could start reaping worthwhile amounts of energy.

Just about all organisms can perform glycolysis. What happens to the pyruvate after glycolysis depends on what kind of organism we are talking about and whether or not free oxygen is available.

Remember that in glycolysis we needed a supply of NAD^+? And that ended up producing a supply of NADH? Could some bright person here figure out a way, given these facts, to get us all the NAD^+ we wanted?

BRIGHT PERSON: Sure, reoxidize the NADH. That would give you NAD^+ again.

FARNSWORTH: Ah, all well and good, but aren't you going to have a little souvenir left over?

BRIGHT PERSON: Uh, yes, I guess. You'd have an H left over.

FARNSWORTH: Quite so, a hydrogen atom. And what you do with that hydrogen represents one of the great biochemical divisions between groups of organisms. If you take that hydrogen and donate it to an organic molecule, in the absence of oxygen, the process is called *fermentation*. This is an example of *anaerobic respiration* and is characteristic of single-cell organisms such as bacteria and yeasts, which are microscopic fungi related to mushrooms. Under special circumstances you can have fermentation in higher organisms, even including people. If the hydrogen is donated to an oxygen to make water, we then have *oxidative*, or *aerobic respiration*, which will be the topic of the next lecture, but let's finish up by looking at fermentation in a little more detail.

Fermentations of different kinds are enormously important biologically and in human terms. For example, consider the alcoholic fermentation performed by yeasts. Yeasts are microscopic organisms, which handle the problem of taking care of the surplus hydrogen from the oxidation of NADH. First, a yeast grabs a pyruvate molecule and then splits it into a molecule of carbon dioxide (which it dumps into the atmosphere) and a 2-carbon molecule called *acetaldehyde*. The acetaldehyde picks up the extra hydrogen, and what's left is ethanol, good ol' ethyl alcohol. Beer alcohol. Wine alcohol. By the way, does anybody here know what percent alcohol wine is?

STUDENT: 12%?

FARNSWORTH: Good shot. A naturally produced wine runs about 9–14% alcohol. Some wines, like sherry and port, have distilled brandy added to jack the alcohol content up to 20% or so. Incidentally, *why* is the alcohol level about 12%?

STUDENT: Because when the yeast starts to break down the sugar in the grape juice, it runs out of sugar—that's why table wines like burgundy don't taste sweet.

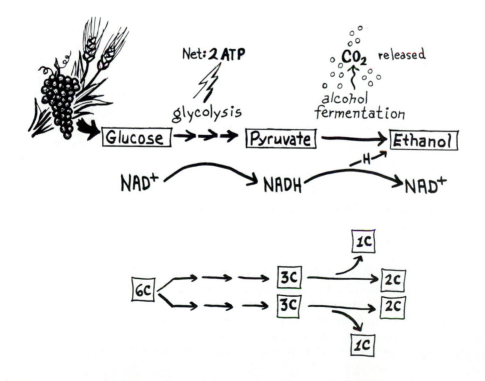

FARNSWORTH: Good thought. Wrong, but an excellent hypothesis nevertheless. Think for a second. What do you put on a cut to disinfect it?

THE STUDENT: Iodine.

FARNSWORTH: What else?

THE STUDENT: Alcohol?

FARNSWORTH: Alcohol! And what does the alcohol do?

THE STUDENT: Kills the germs.

FARNSWORTH: And a germ is?...

THE STUDENT: Yeah, like a bacterium or a yeast.

FARNSWORTH: Right. The reason the wine is about 12% alcohol is that at about that concentration, the yeast is killed by the alcohol. While we're on the subject, the other product of alcohol fermentation is carbon dioxide. If the fermentation takes place in a closed vessel, like a bottle, the CO_2 is dissolved in the wine and creates pressure in the bottle. When we open the bottle, surprise! Champagne. When you make bread, you add yeast organisms to the flour and sugar. The yeast manufactures carbon dioxide, which makes the bread rise. The alcohol is driven off by the heat of baking. You could get drunk, I suppose, by eating a lot of raw bread dough, but I don't recommend it.

So alcohol fermentation is obviously important. Bacteria do other fermentations that produce interesting leftover by-products. Vinegar is made when alcohol is fermented by bacteria to produce acetic acid. The rancid-butter smell, butyric acid, is produced by a bacterial fermentation. A lot of important industrial and pharmaceutical products are produced by bacterial fermentations conducted on a huge scale.

There's one last fermentation I want to talk about, and it will be of interest to all you jocks in the class. Normally, muscles operate by oxidative respiration, which we'll talk about next time. But there are times when the demand for energy release is greater than the capacity of the circulatory system to bring oxygen to the muscle cells. There is a shortage of oxygen, in other words. Normally, if a higher organism's cell runs out of oxygen, it shuts down its operation. But there is a kind of "emergencies only" metabolic pathway that lets you get a little energy out of glucose, even without oxygen. What you do is combine pyruvate and NADH and convert it into *lactic acid* and NAD^+. Unfortunately, lactic acid tends to be toxic, and as it starts to build up in the absence of sufficient oxygen, eventually the muscle will stop functioning. This is the point where you just can't do that last bench press, no matter how hard you try. When you stop the physical activity, you're still breathing hard and the incoming oxygen permits the breakdown of the lactic acid into nontoxic products, so there are no permanent effects.

Let's wrap this up now. We talked about energy and the energy transactions involved in making and breaking bonds and in doing chemical reactions. We saw that you needed activation energy to break a bond. Then we talked about ATP and ADP, and how they act sort of like storage batteries. We then moved on to consider

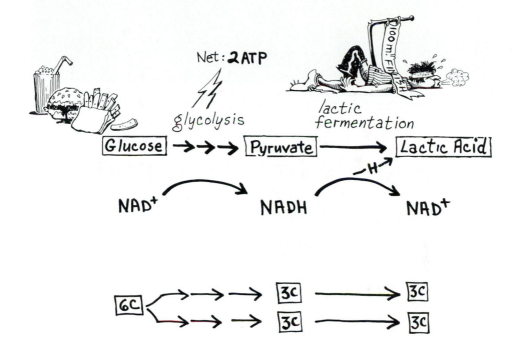

how to crack a carbohydrate molecule apart to get at the bond energy needed for making ATP. We went through glycolysis and fermentation. Next time, we will talk about what happens to pyruvate if you have oxygen present. See you later. (*Students come up to check their notes.*)

Day 5

The Citric Acid Cycle and Electron Transport

Professor Farnsworth enters wearing a Red Sox sweatshirt and running shoes.

FARNSWORTH: Good morning. Somebody, Satchel Paige, I think, said "Never look back. You don't know what might be gaining on you." How true. Let me catch my breath (*he pauses for a few seconds*). Phew! Okay, I'm ready; let's go. What were we talking about last time? Anybody?

STUDENT: Energy.

FARNSWORTH: Fine. I obviously could use some. What about energy?

SAME STUDENT: You talked about glycolysis and how you could produce a couple of ATP's when you broke the glucose down into pyruvate. Then you talked about fermentation and how microorganisms and, under special conditions, animals could break down the pyruvate without oxygen to give you some ATP's, CO_2, and different kinds of organic molecules, like ethyl alcohol or acetic acid.

FARNSWORTH: Good. And what did I promise you for today? Somebody else.

ANOTHER STUDENT: What happens when you break down the pyruvate if you *have* oxygen. I think you called it "aerobic respiration."

FARNSWORTH: Fine, but it will be more than just a recitation about a process. Today, you will learn about something that is truly beautiful. Ideas can be beautiful, as well as things. You can have a gorgeous painting or an ugly painting. You can have a beautiful idea or an ugly idea. A beautiful idea has balance and symmetry and is an elegant way of going about something. You will be stunned when you see how pretty this idea is, and you will be amazed at the cleverness of the people who figured it out.

Now, I have to give you the same warning I gave you before last lecture. I'm going to leave out a lot of steps in this process. Go back to your book for the detail.

Aerobic respiration. We are going to start with pyruvate, remember. That's what we ended up with last time, but this time we will end up with some ATP's—quite a few ATP's, as a matter of fact—and a couple of simple compounds that are not broken down further by the cell. These compounds are carbon dioxide and water.

These represent by-products—not waste products. A waste is something you don't use at all in a process, like the rock part of gold ore; a by-product is something that is left over after you finish your process, like the exhaust that comes out of your car's tailpipe. The carbon dioxide is just dumped—it diffuses out of the cell and into the circulatory system and eventually gets to the atmosphere. Humans simply breathe it out. The by-product water can be used if the cell needs it, or if the cell has enough water, it can be dumped—breathed out, sweated out, or peed out.

The fun starts because we don't go directly, in a straight line, between pyruvate and the by-products. There is something marvelous, and totally unexpected, in between, and that thing is a chemical wheel, or cycle. Let's see now how it works, a step at a time.

(*He puts a transparency on the projector.*) Let's start with the pyruvate. Remember, now, how many molecules of pyruvate do we get from one molecule of glucose? Right, two. And how many carbons does each pyruvate have? Again right, three. Let's first do something with the pyruvate molecule.

What we do first is oxidize it. Remember, *oxidize* doesn't mean add oxygen; it means that the thing being oxidized donates electrons to something else. An *oxidizing agent* picks up the electrons. Like a secret agent picking up the missile plans. I wish there were a different term than *oxidize* in common use for this process, but there isn't, so we'll have to live with it. Anyway, when we oxidize the pyruvate, something rather dramatic happens. The pyruvate molecule yields a couple of electrons and a tagalong hydrogen, all of which we use to reduce NAD^+ (remember good ol' NAD^+) into NADH. At the same time, we strip off a CO_2, which leaves us with a molecule with how many carbons? Correct, two. This 2-

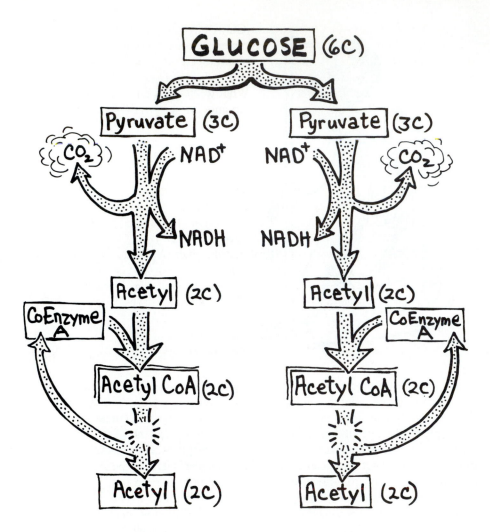

carbon molecule is a functional group, called an *acetyl group*. We then immediately combine this acetyl group with a kind of molecule called a *carrier molecule*, because its function is to carry an important molecule or group from one molecule to another. This particular carrier is called *coenzyme A*. The combined acetyl group and the coenzyme A are called, you guessed it, *acetyl coenzyme A*, or *acetyl CoA*.

Let's stop and see where we are. We started with a 6-carbon glucose and split it into two 3-carbon pyruvates; we then took each of the pyruvates and split off a carbon, leaving us with an NADH and a 2-carbon acetyl CoA per pyruvate. The NADH, remember, has a couple of electrons which we will eventually be able to use to make ATP's. We will return in a little while to what happens to the NADH. What I'm more concerned with is what happens to the acetyl CoA.

Acetyl CoA is a very important compound, if you take my meaning—my meaning being that you'd best be rather familiar with it come exam time. It can be produced by the ***decarboxylation***—the removal of a CO_2—of pyruvate. You can also get it by the breakdown of fats or proteins. It is in some ways a "crossroads" compound because it is the point at which the *real* extraction of energy begins, regardless of the source of the energy-containing compound.

All right, here we have this acetyl CoA. What we want to do first is strip off the coenzyme A. It can then go back and pick up another acetyl group coming out of glycolysis.

Now we are going to do a magic trick. We are going to make the constituents of the acetyl group disappear. Watch carefully, at no time do my fingers leave my hands. I'm going to get rid of the acetyl but leave lots of protons (H^+) and electrons that I can use to make ATP's. Draw this exactly the way I do in your notes—you're gonna love this. The first thing I do is combine my acetyl with a 4-carbon organic acid, ***oxaloacetic acid***. In answer to your unasked question, yes, you have to re-member these names, and the number of carbons in each, but you don't have to know their structural formulas. This combination gives us a 6-carbon compound called ***citric acid***. Yup, orange juice citric acid.

Remember now, I'm leaving out a lot of steps, but showing you the transition steps, where something important happens. We take our 6-carbon citric acid and pull off a carbon dioxide, which disappears; we won't see it again. We pull off a couple of electrons and a hydrogen and reduce an NAD^+ to NADH. What we are left with is a 5-carbon compound called α-***ketoglutaric acid***. Let me stop here for a sec. Any questions? Yes, Delts T-shirt.

DELTS T-SHIRT: You've been talking about starting with an acetyl CoA. What happened to the other one? I thought you got two acetyl CoAs out of the glucose.

FARNSWORTH: Exactly. You *do* get two. I'm showing you only what's happening to one of them. Exactly the same thing is happening to the other one. For example, I showed only one molecule of CO_2 being pulled off the citric acid, but since there is *another* citric acid that was produced by the other acetyl group, you really have 2 CO_2 molecules liberated at this step—one from one acetyl, the other from the other. I'll show you only one, though. Okay, may I continue?

We now take our 5-carbon α-ketoglutaric acid and jigger around with it. We pull off another CO_2. We reduce an NAD^+ to NADH. And because we're breaking a lot of bonds in this step, we can synthesize an ATP from an ADP and phosphate

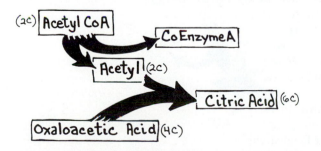

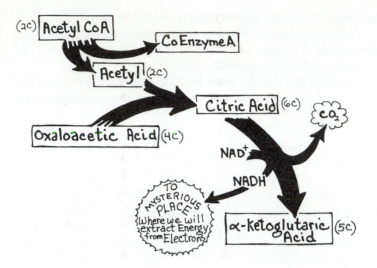

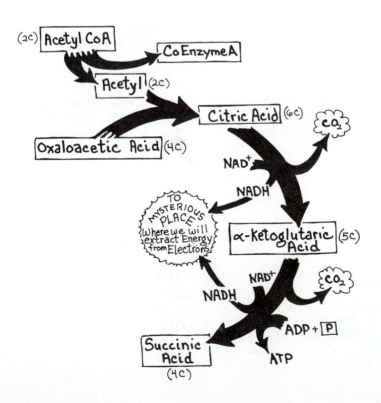

group. This is called ***substrate-level phosphorylation***. A lot going on. What we are left with is a 4-carbon compound called *succinic acid*.

The next step is kind of interesting. We have this 4-carbon succinic acid, right? Well, we kind of rearrange this molecule in such a way that one of the bonds holding it together is changed from what's called a *single bond* to a *double bond*. In the process, we can pull out a hydrogen and a couple of electrons. Now, so far, what have we done with this hydrogen and its electrons?

STUDENT: Reduced an NAD$^+$ to an NADH.

FARNSWORTH: Right, but now, because of the way we got the hydrogen and its accompanying electrons out, we can't use an NAD$^+$. We use another electron carrier molecule called ***flavin adenine nucleotide***. Everybody repeat that after me. *Flavin adenine nucleotide*. Good, you can pronounce it. Useless knowledge because everybody calls it *FAD;* the letters are pronounced separately, rather than as the word *fad*. FAD can pick up two hydrogens when it is reduced, so what we end up with when we rearrange our 4-carbon succinic acid is another 4-carbon acid called *fumaric acid* and an FADH$_2$.

Patience; we're almost done. We now add a water to our 4-carbon fumaric acid, not to dissolve it, but to donate the hydroxyl group in the water to the fumaric acid. When we've done this, we have another 4-carbon acid, *malic acid*.

Now, watch carefully, the hand is quicker than the eye. I break a little bond, so,

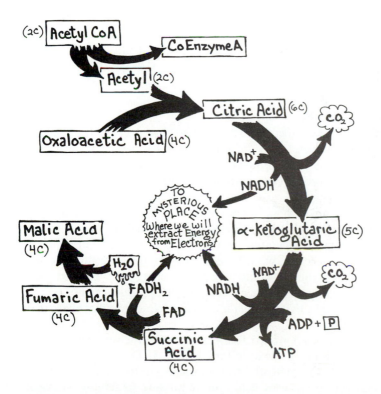

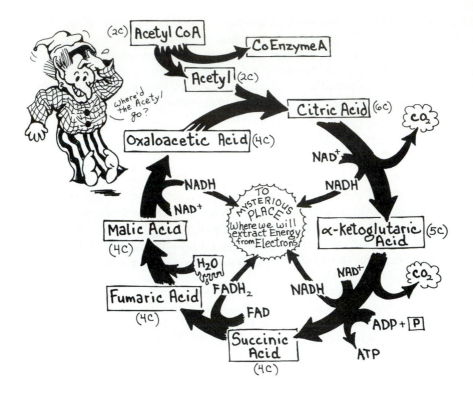

and pull off a hydrogen and a couple of electrons which I give to my beautiful assistant holding an NAD^+, which she reduces to NADH, and PRESTO CHANGE-O! I now have oxaloacetic acid, which is what we started with, and the *acetyl group has completely disappeared! Ta DA! (He does an elaborate sweeping stage bow. The class bursts into spontaneous applause.)*

Thank you, thank you. Now I'm going to do what no good professional magician ever does; I'm going to show you how I did the trick. What we really want to do in this whole process is pull as many protons and electrons out of the original glucose molecule as we can because we can use these, in a way I'll describe in a couple of minutes, to make ATP's.

The acetyl group we started with has plenty of these protons and electrons, but to pry them out, we have to, as it were, prepare the acetyl molecule. We do this by making it into citric acid when we combine it with oxaloacetic acid. This is where the beautiful part comes in. Clearly, we will need a big supply of oxaloacetic acid if we have a lot of acetyl molecules. It would take a lot of energy to synthesize these oxaloacetic acids, but by having this cyclic process, the end product can be recycled over and over, at very little cost. Essentially it is a chain of equations that chase their tails.

So what happened to the acetyl group? Remember, along the way, we liberated 2 carbon dioxides? The formula of the acetyl group is C_2H_3O. You can immediately see that if you're going to give off 2 CO_2 molecules in the cycle, and you're going to start and end with oxaloacetic acid, you're missing something—oxygen.

Where does it come from? From oxygen in the air? No. It comes from water. I add water at the appropriate place, pull it apart to get the oxygen, and use the leftover hydrogen from the water to fill vacant bond locations where I pulled off my carbons.

Thus, you really need oxaloacetic acid, acetyl CoA, and water to make this *citric acid cycle* work. The citric acid cycle is sometimes called the *Krebs cycle,* after Sir Hans Krebs, the British genius who first worked out the details.

After the citric acid wheel has made one turn, and processed 1 acetyl molecule, what we have are 1 ATP, 3 NADH, and 1 $FADH_2$, each containing hydrogen and electrons picked up during the Krebs cycle. Remember that a molecule of glucose yields 2 acetyl molecules, so after the 2 acetyls each pass through a citric acid cycle, there are a total of 2 ATP, 6 NADH, and 2 $FADH_2$ molecules. What we want to do now is see how the energy contained in those electrons can be used to make ATP's.

Before we can talk about the process, I have to tell you a little about where this is all taking place. The process of glycolysis can take place almost anywhere in the cell, but the enzymes for aerobic respiration are contained in structures called *mitochondria,* which are scattered inside the cell.

MITOCHONDRION

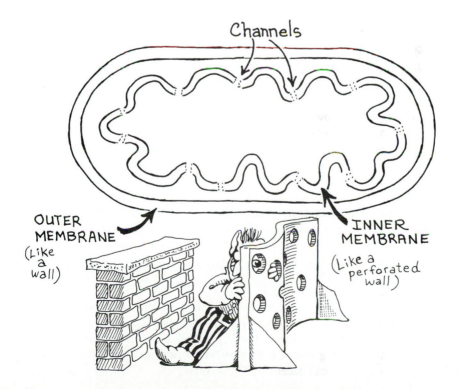

Channels

OUTER MEMBRANE
(Like a wall)

INNER MEMBRANE
(Like a perforated wall)

A *mitochondrion* is shaped sort of like a vitamin capsule. In its construction, you can think of it as a kind of fortress. The outer shell is more or less like a solid wall. Inside this wall, or *outer membrane,* there is an inner, porous wall, or *inner membrane,* which is penetrated by holes, or channels, sort of like the gun ports in a fortress. Technically speaking, these structures are called *membranes,* as I've said, but I've called them "walls" to give you the idea that there is some thickness to them. Don't get these membranes confused with the cell wall of plants, which is outside the whole cell.

Anyway, with these two membranes, the mitochondrion is divided into two fluid-filled compartments, separated by the inner membrane. Glycolysis takes place outside the mitochondrion. The pyruvic acid it produces finds its way into the inner compartment, where it is converted to acetyl CoA, which in turn passes through the citric acid cycle, yielding NADH and $FADH_2$.

The outer membrane is kind of boring—all the action takes place in this hole-filled inner membrane. These holes, or channels, are not just smooth things like the inside of a pipe—they're lined with all sorts of compounds that do amazing things.

Okay, fasten your seat belts, here we go. We'll take our NADH and $FADH_2$ molecules to the *inside* of the *inner* membrane, and here is the first critical event. We are going to take the hydrogens and electrons off the NADH and $FADH_2$ and *separate them.* Different things are going to happen to them. This is a place where lots of people get confused. Keep it straight that one thing is happening to the electrons, and something different is happening to the hydrogens. But wait a minute! I said we are going to take the hydrogen and electrons off the NADH. What do you think happens to the *NAD* part? Guess.

STUDENT: It gets changed back to NAD^+.

FARNSWORTH: Right! It can now go back to the citric acid cycle and pick up more hydrogens and electrons. Economy, see—you don't have to make new NAD^+ all the time.

Okay, now we have our hydrogens—let's look at the hydrogens first—sitting there at the entrance to the channels. Remember diffusion from high school biology? Here you have holes in a membrane. What would you expect to happen to the concentration of hydrogens on both sides of this inner membrane if you just went away for a while and came back later? (*Silence*) C'mon, try! You can't have forgotten this quickly. Lots of hydrogens on the inner side of the membrane, holes through the membrane, no hydrogens on the other side. Try again, what happens after a while? Ah, good, the student with the—what is that—chartreuse-colored blouse in the back middle?

CHARTREUSE BLOUSE: Well, I'm not sure, maybe if you waited long enough, there would be the same number of hydrogens on both sides of the membrane?

FARNSWORTH: Perfect! The hydrogens would diffuse through the holes and after a while would reach equilibrium—the concentration of hydrogens outside would be the same as the concentration inside. *Initially* you had a greater concentration on the inside as your starting condition because we were bringing the hydrogens to the

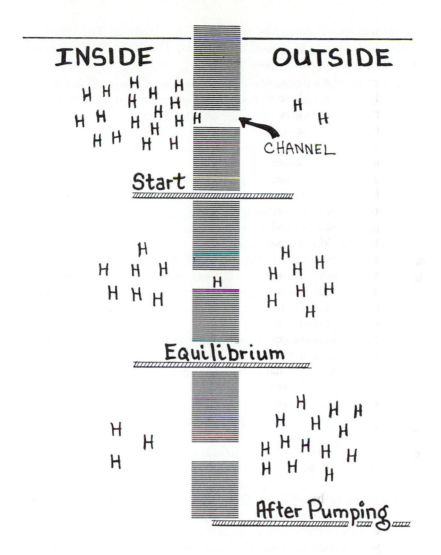

inside of the membrane after picking them up in the citric acid cycle. After diffusion through the membrane took place, the concentration would equilibrate. Now, let's hold on here. Suppose, just suppose, that for some reason I *wanted* to end up with *more hydrogens on the outside than on the inside.* Could I do it? Could I produce a condition where I started with more hydrogens on the inside and ended up with more on the outside? You must know me well enough by now to know that the answer is "yes." But how? *How* could you produce this higher concentration on the outside? I'll tell you how. By spending energy. You could *pump* those hydrogens through the membrane, as many as you had energy for. So if you had a source of energy, and you really wanted most of those hydrogens on the outside, you could do it. But now we have a problem. Where are we going to get the ener—

STUDENT: (*Jumping up and down*)The electrons! The electrons! You could use the energy in the electrons you pulled off the NADH. They're in the right place, too, right on the inside of the membrane.

FARNSWORTH: Ha! Mebbe we done got ourselves a scientist here. You folks sitting around that guy, be sure to copy his notes; he's paying attention. Yes indeed, you could, and do, use those electrons to pump out the protons. The way it's done is kind of interesting.

I didn't tell you this before, but electrons can exist at different energy levels. If you remove a little energy from the electron, it slips down to a lower energy level. The electrons arriving from the citric acid cycle are at a very high energy level. What we do then is pass these electrons on through a series of electron-transport molecules. Each time we pass the electron on to the next transport molecule, the electron loses a little energy, sort of like a leaky bucket. Guess what we use that energy for? Of course; we pump protons out.

The first of the carrier molecules is called **NADH dehydrogenase**. In turn, the electrons pass to **coenzyme Q**, **cytochrome b**, **cytochrome** c_1, **cytochrome c**, **cytochrome a**, and finally **cytochrome** a_3. Cytochrome a_3 takes 4 electrons, picks up 4 hydrogens on the outside of the inner membrane, and uses the electrons and the hydrogens to reduce a molecule of oxygen (O_2) to 2 molecules of water.

This whole system of carriers is called the **electron-transport chain**, and it functions to provide energy to operate the proton pump. Now, the more curious of you will want to know *exactly* how the energy is used to drive the proton pump. Well, I have good news for you, in two respects. *Nobody* knows exactly how the proton pump works. The reactions work so fast that we can't tell what's happening. So that means one less thing for you to memorize, and a potential Nobel prize for whichever one of you figures it out.

So now we have all those protons on the outside of the inner membrane, and we use some of them, along with the electrons and free oxygen, to form water, which is our final stable compound, the end product of respiration. But haven't we forgotten something? WHY DO WE WANT ALL THOSE PROTONS OUT THERE? What is the point of this whole respiration exercise? To use the energy of food molecules to make ATP's, and so far, we've made only a handful of them.

Consider this, though. We have used a lot of energy to get all those protons outside the membrane, haven't we? We now have a high concentration of protons outside, relative to the inside. What is going to happen to all those protons?

STUDENT: They're going to want to get back inside.

FARNSWORTH: Well, maybe "want" isn't exactly right, but that's the general idea. Because there's a higher concentration outside than in, a *diffusion gradient* is created, such that a lot of those protons are in effect going to be *forced* back through the membrane. Now, can you guess what's going to happen when those protons are pushed back into the membrane? You got it, the energy contained in the protons that are pushed by the diffusion gradient back into the membrane is used to bind phosphate to ADP to make ATP. In addition to the diffusion gradient, there's

another source of stored energy. You have more positive hydrogens outside the membrane than inside, right? Well, anytime there is a difference in electric charge between one place and another, you have an electrical potential, a voltage. That electrical potential can be used to do work, and here the work is done to make ATP. So, after this incredibly complex and roundabout route, we *finally* get our energy released from the glucose. *Gaaaaaah. (He slumps over the projector, clearly drained by the effort required to explain all this.)*

Sorry, this is a tough explanation. Let's see if we can wrap it up now. The simplest way of showing aerobic respiration of glucose is with this basic equation

$$C_6H_{12}O_6 \text{ (glucose)} + 6O_2 \rightarrow 6CO_2 + 6H_2O$$

Glucose plus oxygen gives you carbon dioxide and water. Going to the next level of complexity, in glycolysis, you have

$$C_6H_{12}O_6 + 2ADP + 2P_i \text{ (inorganic phosphate)} + 2NAD^+ \rightarrow$$

$$2C_3H_3O_3 \text{ (pyruvate)} + 2ATP + 2NADH + 2H^+ + 2H_2O$$

The extra hydrogens and oxygen on the right side of the equation come from disassembling the phosphate group when it's attached to the glucose.

Converting the pyruvate to acetyl CoA, you have

$$\text{Pyruvate} + NAD^+ + CoA \rightarrow \text{acetyl CoA} + NADH + CO_2$$

Plugging in what happens, then, in the citric acid cycle and along the electron-transport chain, our overall reaction for aerobic metabolism of glucose is

$$C_6H_{12}O_6 + 36NAD^+ + 36ADP + 36P_i + 36H^+ + 6O_2 \rightarrow$$

$$6CO_2 + 36ATP + 6H_2O + 36NADH$$

All told, we spent 2 ATP, back in glycolysis, and got 38 back, for a net return of 36 ATP from 1 molecule of glucose.

Okay, a recap now. You can get energy from a variety of molecules: carbohydrates, fats, and proteins. Preliminary processing of these foods takes place in the digestive system; final breakdown takes place in the cells, more specifically, in structures within the cells called *mitochondria*.

You can metabolize food either with or without oxygen, but using oxygen yields far more of the molecules that carry energy around the cell, molecules called *ATP*.

If you start with a carbohydrate called *glucose*, your first step is called *glycolysis* and yields a product called *pyruvate*. If you don't have oxygen, you can ferment the pyruvate and get a few ATP's and some kind of organic by-product. If you do have oxygen, you change the pyruvate into an acetyl molecule, which you can also get by breaking down a protein or a fat.

The acetyl molecule is fed into a process called the *citric acid cycle,* where it is torn apart, yielding carbon dioxide, which is carried away, and a supply of hydrogens and high-energy electrons.

The energy in the electrons is used to pump the hydrogens to the outer side of the inner mitochondrial membrane, where some of the hydrogens are combined with oxygen and the now energy-poor electrons to form water. The surplus hydrogens are pushed back toward the inside of the inner membrane. In their passage through the membrane channels, somehow the energy used to force the hydrogens back into the middle of the mitochondrion is used to bind a phosphate to an ADP, making our ATP. Most of the energy released from the glucose is converted to ATP in this electron-transport proton pump mechanism, in a process called *chemiosmosis.* Any questions?

STUDENT: Yeah, do plants do this? I thought plants burned carbon dioxide for energy?

FARNSWORTH: Ah, an erroneous assumption, but a good point. Plants indeed use carbon dioxide, but not for obtaining energy from food. Plants use oxygen just the same way that animals do to extract energy from their food. What is different is that plants use carbon dioxide to *make* their food, and that is what we will talk about next time.

Day 6

Photosynthesis

Professor Farnsworth enters wearing a beautifully tailored Italian silk suit with a Sulka shirt and handmade English shoes. He is carrying a very large briefcase.

FARNSWORTH: Sorry for the dude clothes, but I have a plane to catch and a man to meet after the lecture. We are now about a third of the way through the very technical lectures at the beginning of the semester. You're being very patient, and I promise you that when we finish with this block of lectures, I'll give you a break and maybe we'll have some fun with something a little out of the ordinary. (*He displays a cryptic smile.*)

For right now, though, we're going to pick up from last time. Fortunately, someone asked about plants and carbon dioxide just before the end of last lecture. Today, we're going to look at a rather marvelous thing plants can do with a little carbon dioxide and water.

For the past two lectures, we've been talking about how to break glucose down to extract the energy contained in its bonds. A reasonable person might ask, "Where does the glucose come from?" Well, if you eat something, the glucose might come from the body of the thing you eat, but what if you don't "eat" in the usual sense? There is a group of organisms called **autotrophs** (*auto* meaning "self" and *troph* meaning "nutrition") that can manufacture their own glucose. They use the energy contained in light to assemble carbon dioxide and water into glucose. The organisms you're most familiar with that do this are the green plants, and the process by which they do it is called **photosynthesis**, from *photo* meaning "light" and *synthesis* meaning...?

STUDENT: "Putting together."

FARNSWORTH: Good, you're awake. Just checking. If you just looked at the overall reaction, you would say that photosynthesis is simply the reverse of aerobic respiration. Here's the equation for respiration

$$C_6H_{12}O_6 + 6O_2 \rightarrow 6CO_2 + 6H_2O + \text{Energy (ATP)}$$

and here is the equation for photosynthesis

$$CO_2 + 6H_2O + \text{energy (light)} \rightarrow C_6H_{12}O_6 + 6O_2$$

90

It is true that the start and end points are the same, but photosynthesis is not just respiration run backward. We are really going to have to look at it almost as if it were a totally new process that we don't know anything about.

Before we can get into the chemistry of photosynthesis, we have to talk a little bit about physics first. (*Groans from class. Professor Farnsworth does a stage double take and switches over to a fake Nazi accent.*) Zo! You do not like physics, eh? I haf vays of dealing mit dat. Maybe next I vill gif you CALCULUS if you do not like mine physics, eh? (*Switches back to normal voice*) Okay, okay, only a *little* physics, all right? Photons. Just photons. What's a ***photon***? (*Rhetorically*) I wish you hadn't asked me that. Sometimes it's one thing; sometimes it's another, depending on how you look at it. Why can't I just tell you what it is? Because it is something called an *intangible*, something that may not have a physical existence as we know it, but it has properties that can be seen or measured. For example, can you kill somebody with a bullet? Of course. A bullet is tangible; it has weight, length, color, smell. On the other hand, can you kill somebody with hate? Can you describe the effects of hate? Are the effects of hate different from the effects of love? Can you predict what will happen if someone hates somebody else? Yes, to all the above. But have you ever seen hate? Is it heavy? Does it glow in the dark? Negative to the above. Hate is something that no one has seen, something no one will ever see, but everyone knows what it is and what it does. It is an intangible.

A photon is like that. It produces effects that we can see and measure quite easily, but the photon itself defies precise description. You can't even call it a thing, because *thing* implies some physical existence and we're not even sure about that.

It is much easier to describe the properties of photons than it is to describe the photons themselves. Under certain conditions, photons behave like tiny physical particles, like subatomic particles. Something that gives off light energy gives off photons that act like particles when they hit things. On the other hand, when the photons pass through materials like glass, they have properties that suggest a wave-like nature.

Now, I had a lot of trouble trying to think about something that could be a particle *and* a wave. *Particle* isn't hard—bullets coming out of a machine gun. *Waves* aren't hard. Drop a rock in a puddle, and waves spread out from it. But particle *and* wave...?

I came up with a way of thinking about photons that is helpful in looking at the properties of these little whatever-they-ares. Mind you, I'm not saying this is the way they actually are—nobody knows that—but you can look at them this way. Look at them as waves *of* particles. Imagine half a dozen M60 machine guns pointed at a target. They're hooked together so that they all fire their bullets together, dum-dum-dum-dum, like that. Picture all the bullets going to the target. Notice something, eh? The bullets go in *waves*. Clever, huh? What happens if we speed up the rate of fire of the guns? The distance between the waves of bullets gets shorter, or the *wavelength* gets shorter, right? And if this wavelength gets shorter, what happens to the *frequency*, or number of bullets that hit the target in a given time period? It goes up. The shorter the wavelength, the higher the frequency. Just like light energy.

Now, does a bullet have energy? Sure does. Are you going to have more or less energy hitting the target with short-wavelength waves of bullets than with long-wavelength ones? Clearly, with short wavelengths—there are more bullets hitting the target in a given time period.

Light is given off in waves, and the shorter the wavelength of light, the more light energy that hits the target. The difference between the photons and the bullets is that the energy of each bullet is constant—it is their behavior in groups of bullets that changes with wavelength. With photons of light, the shorter the wavelength, the more energy each photon contains. The net effect is the same—the shorter the wavelength, the more energy available for work.

In a nutshell, red light has a long wavelength, blue has a short wavelength, and everything else visible is in between. Plants are green, so that means they absorb green light, right? (*Silence*) Ack! You're *not* awake. The plant *reflects* green light. What it absorbs is red and blue light. Why do you suppose the plant doesn't absorb green light to use for photosynthesis? Because there's very little green light that gets through the atmosphere. The atmosphere absorbs most of the green light coming from the sun. Why is the sky blue at midday? Because mostly blue light gets through. Why is the sky

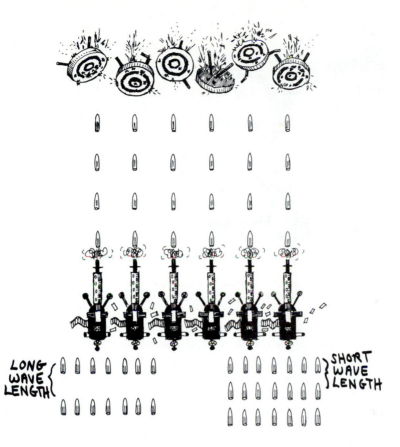

red at sunset? Because at sunset, when sunlight is grazing the atmosphere instead of passing straight down through it, the dust in the atmosphere absorbs most of the blue, leaving red and what little green there is. So if a plant is going to use sunlight passing through the atmosphere, it is going to have to be specialized to use the wavelengths of light that are mostly there, that is, the red and the blue.

So where are we now? A plant uses the energy contained in photons of red and blue light to make sugar out of water and carbon dioxide. Instead of starting at the beginning of the story of exactly how this is done, we're going to start at the middle and get a quick general picture of what's going on.

Our glucose is actually made out of the CO_2. To make glucose out of CO_2, we have to change the CO_2 first into something called *3-phosphoglycerate*, or *3PG*. To do that, the CO_2 is combined with *ribulose biphosphate*, or **RuBP**. Ribulose biphosphate is a 5-carbon sugar that has a couple of phosphates attached to it. From somewhere, then, we're going to need a supply of RuBP. We could, of course, manufacture up a fresh supply of RuBP every time we needed it, but how about something else, something more economical. Do you suppose we could use a...(*pauses expectantly*)?

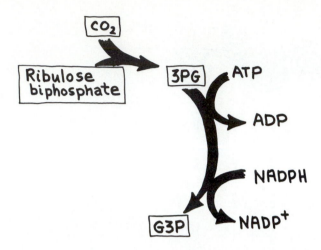

A VOICE FROM THE BACK: Cycle!

FARNSWORTH: Indeed we could use a cycle, just like the citric acid cycle, except this one is called the **Calvin-Benson cycle**, after the people who worked it out.

But we're getting a little ahead of ourselves. We have this 3PG, and we next want to reduce it to something called **glyceraldehyde-3-phosphate**, or **G3P**. To do that requires both energy, in the form of ATP, and hydrogens provided by a reducing agent. We've seen reducing agents before. Remember NAD, of NAD^+ and NADH infamy? In the Krebs cycle? Well, we use a very close relative of NAD here in the Calvin-Benson cycle. It is called **NADP**—just NAD with a phosphate added—and it exists in the $NADP^+$ and NADPH forms.

Since you are already starting to be able to think ahead of me, you will doubtless be able to predict that after we have produced our glyceraldehyde-3-phosphate, somehow we are going to be able to regenerate this RuBP stuff that we used to combine with CO_2 back in the early steps of the process. And you would be right, you *do* do that. But— and this is an important *but*—there is a big difference here between the Krebs cycle and the Calvin-Benson cycle. In the Krebs cycle, we started with one molecule of acetyl CoA, combined it with one molecule of oxaloacetic acid, ran the product, citric acid, through the cycle, and ended up with one molecule of oxaloacetic acid.

The Calvin-Benson cycle is a little more complicated. To get a molecule of glucose, you start with $6 CO_2$, and combine them with 6 RuBP to yield *12* 3PG. You get 12 G3P from the 12 3PG, using 18 ATP and 12 NADPH, but then comes the big difference between the two cycles. In the Calvin-Benson cycle, you pull off two of the 3-carbon G3Ps to make one molecule of glucose. The other 10 molecules of G3P you manipulate, through a long series of steps, to make your 6 molecules of RuBP to run the cycle through again.

Now, as a matter of fact, you don't feed all $6 CO_2$ molecules into the cycle at the same time. You feed them in one at a time, until you have enough G3P molecules to make your glucose. The result is six times around the Calvin-Benson cycle to get one molecule of glucose.

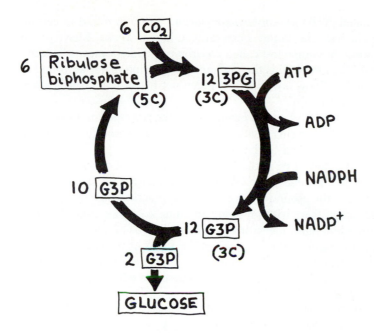

I hope somebody has been alert enough or puzzled enough to notice that I started talking about something at the beginning of this lecture, and then suddenly here we have our glucose, and I haven't said anything at all about the thing I started out with. Anybody, what is that something?

STUDENT: Light.

FARNSWORTH: Of course. I haven't talked about light energy at all. Where does the light come in? Let's think about the raw materials for this Calvin-Benson cycle. CO_2. We can get that from the atmosphere. Ribulose biphosphate. We get that as a by-product of the cycle. What's left? ATP and NADPH. Well, of course, we could get some ATP by metabolizing glucose, but we're trying to *make* glucose here. That would be like saying humans don't have to eat, because they can just metabolize all their carbohydrates, fats, and proteins. No, we don't want to get our ATP for the Calvin-Benson cycle from metabolism of foods. What *could* we use for the energy to synthesize ATP that we can then use to drive the Calvin-Benson cycle?

STUDENT: Light.

FARNSWORTH: Exactly. There are two stages in photosynthesis. The second one, the one we've been talking about, the Calvin-Benson cycle, doesn't use light directly, so it is sometimes called the ***dark cycle***. The first stage, the ***light reaction***, involves the capture of light energy in the form of photons and the use of that energy to manufacture ATP and NADPH. The manufacture of ATP by light energy is called ***photophosphorylation***, and it can proceed in one, or maybe both, of two possible ways.

The first way is called *cyclic photophosphorylation*, and it is found in certain kinds of photosynthetic bacteria. Before I can tell you how it works, I have to tell you about a very important compound called *chlorophyll*. *Chloro* means "green" and *phyll* means "leaf." Chlorophyll is the pigment, or coloring material, found in green plant structures. We know now that it is green not because it absorbs green but because it reflects green light.

Chlorophyll is rather closely related to *hemoglobin*, the red coloring material of blood. It has a large, complicated ring structure with a central magnesium atom and a long hydrocarbon group attached to the rings.

The nice thing about chlorophyll is that it can exist in one of two states, a so-called *excited state*, and a *ground state*. What this means is that if you have a molecule of chlorophyll in the ground state, it is, for all practical purposes, stable. It just sits there, and can sit there without doing anything for a long time. If, however, you blast a chlorophyll molecule with a couple of photons of light, something rather dramatic happens. Some of the electrons in the atoms of the chlorophyll get kicked into a higher energy level, and the chlorophyll is then said to be *excited*. Now, I don't know if "excited" is the right word to describe what happens, but it is the customary one, and we're stuck with it.

An excited chlorophyll molecule is very unstable. If you just left it alone, in less than a millionth of a second the excited molecule would kick out a new photon and the electrons which had originally been boosted to a higher energy level by the photon that hit the molecule would drop back to their original energy level. This process of the release of a photon after electrons are boosted to a higher energy level is called *fluorescence*, and that's how a fluorescent lamp, you know, one of those tube things, works. You coat the inside of the tube with a chemical that can fluoresce, then kick its electrons up to a high level using electricity from the wall plug—you can use electricity as well as light to kick those electrons up. When the chemicals fluoresce, they kick out photons and you can see the light from the tube.

Well, it is very nice that there is this phenomenon called *fluorescence*, but we don't *want* our excited chlorophyll to fluoresce. Who needs plants that glow? What we want to do is move in really fast, before the molecule has a chance to fluoresce, and grab off some of those high-energy electrons and do something with them. Now, to refresh my memory, what do you call something that donates electrons?

STUDENT: A reducing agent.

FARNSWORTH: Right. Excited chlorophyll is a very good reducing agent. We can use excited chlorophyll—I don't know, does that seem funny to you? Excited chlorophyll. I can just see this little green molecule hopping around because it's going to take delivery on a new Corvette. Anyway, excited chlorophyll is used to donate electrons to processes farther down the line in two ways. The first way, as I said before, is cyclic photophosphorylation, and it can make ATP, but it can't make NADPH.

Cyclic photophosphorylation works very much like the electron-transport chain of oxidative phosphorylation. You get your chlorophyll molecule excited, and then before it has a chance to fluoresce, you strip off two electrons. The electrons go first

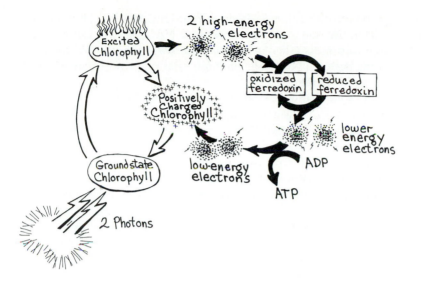

Cyclic Photophosphorylation

to a compound called *ferredoxin*. Ferredoxin can exist as either reduced ferredoxin or oxidized ferredoxin. What we do here is combine our electrons with oxidized ferredoxin, which converts it to reduced ferredoxin. We then pull the electrons off the reduced ferredoxin and pass them to the next oxidized compound in the chain. And what do you suppose happens to the reduced ferredoxin when we pull the electrons off?

STUDENT: It becomes oxidized ferredoxin.

FARNSWORTH: Correct. The oxidized ferredoxin can then pick up another pair of high-energy electrons and start the business over again.

In the meantime, we are passing those electrons down a chain of oxidizing agents, and each time the electron gets passed on, it loses a little energy, which is passed off as heat. It is passed off as heat, *except* during one step in the process, where the energy drop is large enough that we can use the energy to synthesize a molecule of ATP.

At the end of the chain, we pop our pair of electrons out, but now they have lost the energy they picked up when the photons hit the chlorophyll. What do we do with them then? Well, remember the excited chlorophyll? We pulled a couple of high-energy electrons off it, didn't we? That means it was left with a positive charge, and could now absorb some electrons if they were available. Well, now we have a fine, strapping pair of electrons, so we combine them with our positively charged, formerly excited chlorophyll, and what we get is a stable chlorophyll molecule in the ground state, waiting to be hit with another pair of photons and start the whole business over again.

Cyclic photophosphorylation is pretty effective in producing ATP's, but remember, to make the Calvin-Benson cycle work, we need NADPH. We can get both ATP *and* NADPH by a process called ***noncyclic photophosphorylation***, found in green plants and some bacteria. Noncyclic photophosphorylation is more efficient than cyclic photophosphorylation.

Noncyclic photophosphorylation is kind of a neat process. You start with water. Split it up into hydrogen (H^+), oxygen (O_2), and electrons. The oxygen gets dumped to the atmosphere, luckily for us, and the hydrogens we won't worry about for a minute. Let's follow the electrons.

Okay, now remember, in cyclic photophosphorylation you excited the chlorophyll, pulled off those high-energy electrons, passed them down a chain, made an ATP using some of the energy in the electrons, and then restored the chlorophyll

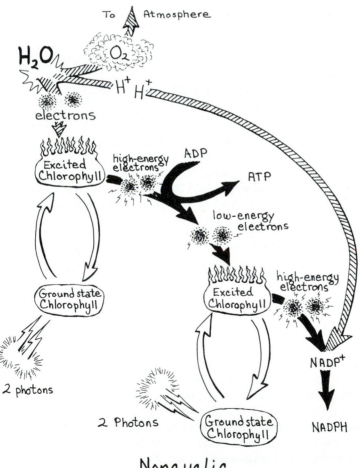

Noncyclic
Photophosphorylation

molecule to its original state by combining it with the electrons that had been spit out of the chain? Okay, in this new process, you take those electrons coming out of the chain and send them someplace else—we'll come to where in just a sec—and the chlorophyll is restored to its ground state with the electrons *from the water you split up*. So, water is feeding electrons into the process—it stops without water.

What do we do with the electrons coming out of the chain? Ah, this is great. We hold on to them for just a bit, and then, watch this. We take *another* chlorophyll molecule, pump it up with photons, strip the high-energy electrons off, and restore it to the ground state with the low-energy electrons we just pulled out of the chain. Now we have a ground state chlorophyll and a couple of high-energy electrons to play with. What do we do with them? We feed them into *another* electron chain, a second one, and we pass them down. But what do we do with the energy we get out of them? Ah, well, remember we had some hydrogens left when we split up the water? Remember that? And what is it we want to get out of this whole mess, the thing that we need for the Calvin-Benson cycle? NADPH. You got it. And how do we get NADPH? By combining $NADP^+$ with, ah, you see it now, hydrogen, using the energy from the high-energy electrons we passed down the chain. The electrons themselves get picked up by the NADPH. So in noncyclic photophosphorylation, we produce both the ATP we need, as electrons are passed through the first chain, and NADPH, at the end of the second chain. Isn't that marvelous? And weren't the people who figured out all these steps clever, especially when you consider that some of the steps take only the tiniest fraction of a second to complete. Now, take a minute to check your notes over with your neighbor and make sure you have everything. (*He steps away from the projector and attends to some busywork on the podium.*)

Okay, let's clean up some details and then wrap this up. We have the general outlines of the photosynthetic process. Where does it take place? In green plants, it takes place in structures called **chloroplasts**. These chloroplasts are structured remarkably like mitochondria, which you will remember from oxidative phosphorylation. The chloroplasts even have inner and outer membranes like mitochondria. The inner membrane is called the **thylakoid membrane**. Photosynthesis involves electron-transport chains and the production of ATP, so we shouldn't be too surprised to find that the ATP is produced by a chemiosmotic process across the membrane, just as in oxidative phosphorylation.

We know that you need water and CO_2 to produce glucose by photosynthesis. But there are some plants that have a peculiar problem. They live in hot environments. Now, the way CO_2 gets into the leaf is through little pores called **stomata**. These pores can be opened and closed. When they're open, CO_2 can diffuse in from the atmosphere. Unfortunately, water can diffuse out, so to prevent themselves from dehydrating, these plants frequently shut their stomata. CO_2 can't get in, and photosynthesis soon eats up the available supply of CO_2 within the leaf. So these guys need to be more efficient at processing the available supply of CO_2 than does a regular plant.

Normally, CO_2 is combined with RuBP, and the product, what was it? (*Silence*) Okay, hint, starts with a *3*.

STUDENT: 3-phosphoglycerate.

FARNSWORTH: Right. 3PG. 3PG is then made into your glucose. The problem is, RuBP is not really that terrific at picking up CO_2 molecules, and really works only at fairly high concentrations of CO_2. So what these special plants do is make a 3-carbon compound called **phosphoenolpyruvate**, or **PEP**, which is a regular CO_2 sponge. It just *sucks* up any loose CO_2 and makes it into an old friend—our good buddy oxaloacetic acid—that we saw in the Krebs cycle. Oxaloacetic acid has 4 carbons, so these kinds of plants are called C_4 plants—sugar cane is an example, as are corn and crabgrass. Now here's a weird thing. This oxaloacetic acid is synthesized in one kind of a cell, called a **mesophyll cell**. The oxaloacetic acid gets changed to malic acid (still 4 carbons) then diffuses to *another* kind of cell called a **bundle-sheath cell**. The CO_2 gets stripped off the malic acid, leaving a 3-carbon compound which eventually gets made into PEP and is recycled. The CO_2 gets fed into the regular Calvin-Benson cycle. These C_4 plants thus have two big advantages in their hot environments: they have a CO_2 sponge (PEP) that grabs lots of CO_2 during the

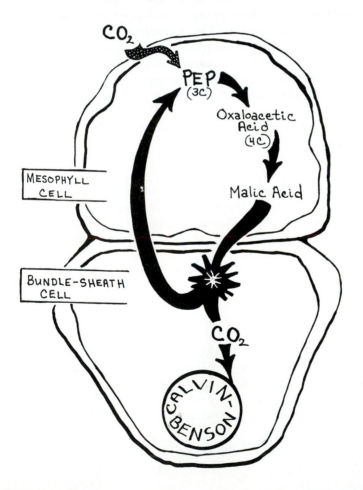

brief periods the stomata stay open, and they have a means of moving CO_2 away from the openings (so it doesn't inhibit PEP) and concentrating it in the bundle-sheath cells so that the finicky RuBP molecules don't know there's a shortage going on out there. The whole process is called C_4 *photosynthesis*. The equivalent process in a regular plant is called C_3 *photosynthesis*, because the first product that's made when CO_2 is incorporated, 3PG, has three carbons.

Okay, what did we talk about today? Photosynthesis is a process that works to make glucose out of CO_2 and water. The water is used as a source of hydrogens and electrons. The CO_2 is used as a carbon source. Light energy is used to produce ATP and a compound, NADPH, which is needed to convert the CO_2 into glucose.

CO_2, which is picked up when it diffuses in through openings in the leaf, is combined with a compound called *ribulose biphosphate* to form 3PG. Using the hydrogens provided by NADPH, and ATP's, the 3PG is changed to glyceraldehyde-3-phosphate, or G3P, which is then made into glucose. Some of the G3P molecules that are produced as a result of this last step are recycled back into RuBP.

I don't know about you, but I think this is a good day's work. Now, unless I run into some unforeseen problems in Marrakech, I should be back on Monday, and I will tell you a story so incredible and beautiful it will have you gasping in amazement. Sorry to rush off now, but I promise I'll make extra time for questions next lecture. (*He hurries out of the auditorium. At the door, a man carrying a briefcase chained to his wrist meets him, and they both rush off.*)

Interview with Professor Farnsworth

The setting is inside Professor Farnsworth's office. It is chaotic and disorderly, looking as though a pack of very curious monkeys had decided to rearrange every physical object in the room. The walls are covered with pictures: Professor Farnsworth in climbing gear, the professor with a tribal chief from Borneo, World War II airplanes, very old daguerreotype photos showing loggers standing atop the trunk of a huge cut-down redwood. There is hardly a square inch of wall visible. A scale model of HMS Beagle, *the British Navy ship that carried Charles Darwin on his voyage of evolutionary discovery, is nestled in the bookcase, which is otherwise filled to bursting with books of every size, age, and subject. There seems to be no topical order to the books; philosophy next to physiology, biochemistry next to Belgian locomotives. The professor's desk is piled high with a clutter of papers and oddities from his trips. An object looking disturbingly like a shrunken human head stares with sightless eyes at the visitor's chair. A student, about 20 years old, sits in that chair. She has been assigned by the student newspaper to do a profile of Professor Farnsworth and get his suggestions on survival for first-year students. The professor, clearly relishing the experience, leans back in the carved-teak reclining chair and takes a big swig of coffee from his ever present cup. She speaks first.*

INTERVIEWER: Dr. Farnsworth, before I begin, I hope you don't mind if I turn on my recorder. I want to be sure that I get everything accurately.

FARNSWORTH: No problem.

INTERVIEWER: Thank you. As I said on the phone, because you, like, have a reputation for being concerned with the problems of first-year students, and because your lectures are sometimes a little unusual, at the newspaper we thought students might like to know more about you, and we wanted to see if you had any ideas that would help people get adjusted to college, if that's okay? (*Professor Farnsworth nods "okay."*) In that case, how long have you been a teacher?

FARNSWORTH: I think almost my whole life. I was constantly getting in fights as a kid because I just *loved* to explain things. I explained things to my parents. I explained things to my dog. I explained things to my electric trains. So in school, when the teacher asked a question, I would just explain. And explain. And explain.

Naturally, the other kids thought I was brownnosing the teacher, for which belief I had to pay dearly at recess. After a while, though, the kids started to find my explanations useful, so I stopped getting stomped in the schoolyard, but *then* I had to worry about not stepping on the teacher's turf. Not easy to be a born teacher when you're 11 years old.

INTERVIEWER: So you're saying you are a born teacher?

FARNSWORTH: Yes, I think so. Now, for heaven's sake, don't turn me into a conceited fathead because I said that. You don't claim credit if you're born with beautiful eyes, or the ability to play music by ear, and it's the same thing with this. It is just there. Every day, when I walk in that door, I am always amazed that somebody pays me to do this. It's like getting paid to drive a Ferrari. Unbelievable. I don't know whether I'm a *good* teacher; that's for somebody else to say. But I've wanted to do this for as long as I've known that there were things to want to do.

INTERVIEWER: So, then, what *is* a good teacher?

FARNSWORTH: You don't waste time, do you? Right to the bone. What is a good teacher? I wish I knew. I don't really think that that is a generally answerable question. A good teacher of what? A good teacher for whom? Let me give you an example. I'm always changing things, trying new things. Every time I do something that is *really* different—you can imagine what it was like the first time I did the biker lecture—there will be some students who come up and tell me that was fabulous, really gave them an insight into whatever it was I was talking about. Then I'll find out through the grapevine that there are others who didn't have the faintest idea what I was trying to do and are very angry that I "wasted" class time. So, to the first group, I'm a brilliant, inspiring teacher. To the second group, who went to the very same lecture, on the very same day, I'm a grandstanding jerk.

That was my first dose of reality therapy when I first started professional teaching. Hoo, boy, was I idealistic. I was going to be such a good teacher that *everyone* in the class was going to just love biology, and work hard, and learn a tremendous amount. Well, as they say, one person's meat is another's poison. My approach was perfect for some of the class, and they thrived on it. For others, it was a disaster.

INTERVIEWER: So why don't you just teach to the middle of the class?

FARNSWORTH: Ah, well, of course you could do that, and it would be perfectly defensible. You could teach in such a way that no one would be offended, and you would do pretty well. No one would be inspired, but on the other hand, no one would be repelled. Middle of the road. But that's where the personality of the teacher comes in, the ultimate variable. I would rather have somebody love my approach and somebody else hate it and have the total come out even, than have everybody indifferent and have the total come out even. That isn't a matter of good or bad; it's just people. Haven't you noticed this yourself? You have some teacher that you think is just wonderful, and you meet somebody who had her two semesters ago, and you start comparing notes, and sure enough, the other person couldn't stand her, for the very reasons that you thought were great. That's why it's so hard to rank teachers. You have to know something about the rank*er*.

INTERVIEWER: Are you saying then that it is impossible to evaluate teaching?

FARNSWORTH: No, not at all. It is very easy for *you* to rank your teachers, according to your personal preferences. The problem comes when you're trying to get some collective, or group, opinion. The student grapevine is notorious for being a lousy source of information about teachers. Unless you know that the person you're asking has the same personal preferences that you do, you might be steered away from a teacher who would be perfect for you, and toward a teacher who will make your life a living hell. Unless you're pretty much like the person that you're asking, I would be very cautious about accepting someone else's recommendations.

INTERVIEWER: That's fine if you're talking about an elective, where you have a choice of teachers. What about required courses, where you're stuck with a particular teacher?

FARNSWORTH: Well, you're really talking about what do you do if the teacher's teaching style and your learning style just don't match. If they do, it is irrelevant that the course is required. If they don't, you go into survival mode. A bad teacher for you can make it almost impossible to develop enthusiasm for the course, and of course, without enthusiasm, it is very difficult to do well, especially in a demanding course for which you have to make sacrifices.

Sometimes, you just have to tough it out and do the best you can. But, in fairness to both the teacher and yourself, at first, give the teacher both a break and the benefit of the doubt. I have seen very, very few deliberately cruel and malicious teachers who "have it in" for a particular student. Oh, the situation exists, to be sure, but it's not common, not common at all. What *is* more usual is a demanding teacher who, to the students, appears to be unreasonably or unfairly demanding. As I said, benefit of the doubt. Maybe the teacher knows something or has experienced something you haven't. My course, Bio 100, is a case in point. I know there are a sizable number of students in that course who think I am an incarnation of Satan himself who makes utterly impossible demands. Well, those poor innocents haven't had comparative anatomy yet, or genetics, and I have. They don't know that they're going to walk into a meat grinder next year. If I can get them mad

enough at me so that they work hard enough to let them coast through the *really* tough courses that will follow Bio 100, I've done my job. If I let them coast through 100, they're gonna get shredded in their next courses. And I'm not just talking about the majors in the course here. We have nonmajors, and it is just as important for them to learn the study skills and techniques in 100 as it is for the majors. They can use that knowledge in their *own* horror courses next year.

INTERVIEWER: That's sort of a patronizing attitude, isn't it? "Do what I say because it's good for you"?

FARNSWORTH: Maybe it is. But maybe that is also part of my job. Students who are, in fact, well prepared in terms of both facts and study attitudes can take an exam to exempt them from Bio 100. For the others, I guess my attitude is something like, "You have come to me and said that you want to be a physician. Very well, I will help you toward your goal, but the form of my help is to make sure that you are well prepared for the formidable demands that will face you."

INTERVIEWER: How do you feel when a student comes to you and says, "I worked as hard as I could, and you gave me a D, and now I'll never be the nurse my dead mother wanted me to be."

FARNSWORTH: Terrible. I really and truly do feel sorry for students who are hard working and just can't seem to meet the requirements. But I also have to think about the patients who might be injured or killed if, out of pity, I let this unqualified person through. Teaching is basically a low-stress job, but there are some tough, tough decisions you have to make. Grading is the part of teaching that most teachers absolutely hate. Incidentally, I've noticed that students invariably say either "I got an A in Bio 100" or "Farnsworth gave me an F in Bio 100." Human nature, I suppose.

INTERVIEWER: Let's talk about that for a while. What are your thoughts on grading?

FARNSWORTH: How many hours of tape do you have? Well, first you have to distinguish between testing and grading. In testing, you're trying to measure what a student learned. In grading, you're trying to rank the student, or put the person in a category with other students who have demonstrated the same degree of learning. Being a good tester, I think, is easier than being a good grader. A good test is one which actually measures what you predict it will measure. For instance, if you tell the students you want them to be prepared to solve problems on an exam, and then give them memory questions, it's a lousy exam. With practice, you can design a fairly decent, reliable, and valid exam.

INTERVIEWER: What's the hardest kind of exam to make up?

FARNSWORTH: Multiple-choice, no question. Let's say you're going to have five possible answers, to reduce the probability of the student's getting the answer by guessing. One of those answers is the correct one, and it's usually not too hard to come up with one or two plausible answers, but to come up with *four* reasonable ones sometimes takes an hour for a single question. Then you have to be careful about things such as asking "Which of the following is the *best* answer?" and then including "none of the above" as one of your answers. Of course, some kinds of multiple-choice questions are trivially easy to make up, like "To which phylum does an earthworm belong?" You have to have some of those, but the real challenges, both for me and for the students, are the problem-style, or thought, questions. You had the course; you must remember some of them.

INTERVIEWER: *(Shudders)* I'll never forget them. I always used to wonder about that. How all five answers seemed to be correct. I never thought about how it might be hard to make them up, but it figures—the harder it is to answer, the harder it must be to write. Okay, grading now. What is your philosophy about grading?

FARNSWORTH: Well, the first thing you have to adapt yourself to, whether you're a student or an instructor, is that no grading system is perfect for all students in all classes. A system that would work to the advantage of one student usually will work to the disadvantage of another. Let me give you an example. Students are always concerned about the "fairness" of a system. But what is "fair"? On the face of it, that would mean that two students who had the same total number of points at the end of the semester would end up with the same letter grade. That's "fair," isn't it? But suppose that that total would have earned both students a B. One of them consistently worked at a B level all semester. The other started out at a C level, got his act together, and by the end of the semester was working at an A+ level. But under this fair system, both of them would get B as a final grade. See what I mean? We could argue all night about this, and there's no real way to settle whether *fair* also means *right*. From a student's standpoint, you're going to run into a dozen different grading systems. What you really need to do, at the beginning of the course, is find out exactly what the instructor's expectations are, and how his or her system works. At a minimum, then, you'll know what you have to do. If the instructor is a little hazy, I think it is legitimate for a student, in a polite way, to press for details.

INTERVIEWER: Okay, another topic. You have a big course, and there are even bigger ones around campus. A lot of students get lost, or feel threatened by these huge courses. As a teacher of a big course, how do you suggest students deal with this?

FARNSWORTH: Well, the first thing to do is realize that being in a big class is really a neutral experience. Nobody is going to volunteer to help you in a direct, personal way, but on the other hand, if you haven't kept up with your reading, nobody is going to reveal that fact and pick on you in class, as might happen, say, in a small language class. Suppose, though, that you *want* some help, advice, attention, or whatever? There are 300 students and 1 instructor. Doesn't that mean that it is impossible to talk to the teacher? In most cases, not at all. Fortunately for you, most students are content to remain invisible and anonymous, so the instructor isn't swamped with students demanding time. A lot of instructors *love* to have students come in and talk to them, because that's the only way they know whether they're reaching the class. Of course, there are always some swine who won't give you the time of day, but they're in the minority—if you have a bad experience with one of those jerks, don't generalize to *all* large-class instructors. Remember, also, that a huge class is really an audience, and the instructor has to be somewhat of an actor to communicate effectively with the class. An actor isn't necessarily the same kind of person as the role being played. A teacher who appears to be a real tyrant in class may not be that way at all on a one-to-one basis. Maybe so, maybe not; the only way you'll know is by going to the office and actually talking to him or her. You'll either have your worst suspicions confirmed, or be pleasantly surprised. In either case, you don't have anything to lose by trying to get to know your instructor. When the two of you are talking, for your teacher the other 698 students in the class don't exist.

You have to remember this, though. Your instructor is a professional, just like a physician. You don't expect to be able to go into a popular physician's office without an appointment and see the person immediately. Your instructor has other courses to teach, maybe grad students, maybe research. My Bio 100 students are always amazed that somewhat less than 20 percent of my working time is spent on 100. A good instructor, like a good physician, will make every effort to see you as soon as possible, but you have to be reasonable about it too, and try to make an appointment as far ahead of time as you can.

INTERVIEWER: Okay, when you do finally get in to talk to an instructor, are there things you *shouldn't* say or talk about?

FARNSWORTH: Definitely. Never ask directly "Will we have to know this for an exam?" Instructors hate to hear that because it suggests that the only thing you're interested in is a minimal effort, and teachers are arrogant enough, or vulnerable enough, to want to think that you *love* their course and want to spend every waking moment on it. Much better to say, "I want to learn as much as I can in Bio 100, but I have only a finite amount of time to spend on each of my courses. Can you suggest how I can decide which topics to put at the top of my list if I have to make priority decisions?" See the difference? You'll get the same answer, but the emphasis is altogether different. Same thing when you go in to ask a question about a grade. If you go in and say, "I wanna know why you gave me a C− on this essay!" the instructor is going to get all defensive and belligerent and you'll end up nowhere. Even if you feel that you were screwed for some reason, start out cool, calm, and reasonable. You're more likely to get a cool, calm, reasonable attitude out of the instructor. Now, I'm not saying you shouldn't stand up for your rights. Absolutely, you should do that. But start out reasonably and rationally, *then* throw, yell, and scream, if you feel that's necessary.

INTERVIEWER: *(Looks at watch)* Oh, I've gotta go to class in a minute, but I have to ask—what is that *thing* there on your desk that looks like a shrunken head?

FARNSWORTH: That? Oh, that's my great-grandfather. He was a gold miner and was killed by the Jivaro Indians in Venezuela. They made him up real nice, don't you think? His partner bought him back from the Jivaros, and he's been in the family ever since. You'd never know it now, but he was a real handsome dude.

INTERVIEWER: (*She looks at him incredulously.*) That isn't true, is it? That's just a monkey head!

FARNSWORTH: (*He tosses the head up and down.*) Anything's possible.

INTERVIEWER: (*A little rattled, she returns to her notes.*) Well, one last question. As one of the more, uh, distinctive teachers on campus, is there anything else you'd like to say about teaching?

FARNSWORTH: Yes. Something I learned in the far east. (*He looks away from the student to a strangely shaped piece of jade resting among some books.*) A teacher is a bridge, over which the student can walk from the past to the future. The wise traveler selects a bridge that is carefully designed, strongly built, and well maintained. If chosen well, the bridge will enable the traveler to cross over barriers impossible to traverse otherwise, and reach distant and marvelous lands traveled only in dreams before.

(*He returns his gaze to the student.*) Thank you for your questions; it's so easy to become preoccupied with the details of teaching that you forget the wonder of it. I hope you won't be late for your next class.

He turns back to his typewriter and is immediately drawn into deep concentration. She gathers her notes into her bag and rushes off to class.

Day 7

DNA and RNA Structure

Professor Farnsworth enters wearing a black burnoose, the hooded cloak of the Arab desert nomads. He obviously fancies himself to be a bit of a Lawrence of Arabia character.

FARNSWORTH: Good morning. I'm happy to report that my trip was successful, but everything you've heard about camels is true. I wore this rig today because I wanted to try an experiment. As you undoubtedly noticed, it is very warm for the season today. It would seem that if you were going to walk around the desert, you would want to wear something white, to reflect the sun, but many of the desert people wear black. Either they're tougher than we already know they are, or there's some property of wearing black clothes we don't know about, so I've got myself wired up with electric thermometers, and after class, I'm going to take a little stroll and see how hot it gets under the burnoose. I wanted to ride a camel across the campus to make the experiment more realistic, but the dean didn't think that was such a good idea.

I told you before I left that I was going to introduce you to something absolutely breathtaking today, and I propose to keep my promise. For the next three lectures or so, I will introduce you to the principles that are at the base of what has been called the *second biological revolution*—the molecular revolution.

What we are going to do is see how it is that a cell, any cell, can do the two most critical things it needs to do during its lifetime. First, since we know that cells divide—and isn't this an interesting thought; cells divide to multiply—there has to be some provision to ensure that each of the two new cells is complete and somehow receives the information that it will need to perform all its cellular duties. When each of these cells, in its turn, divides, it too must pass on to its descendants all the information it inherited from its parent cell. This process of transmission of information from cell generation to cell generation has to be sufficiently foolproof so that it continues to operate for literally millions of generations of cells.

The second thing that the cell has to do is tied very closely to its first main task. Once it has inherited the information it needs from its parent cell, it actually has to use that information to perform a job. The job, in this case, is the manufacture of proteins, many of which are enzymes. Just as a refresher, what's a protein? Yes, Jim. Oh, you're surprised that I know Jim's name, out of the hundreds of students here. Well, no great mystery; Jim comes to my office to ask questions. In high school, we would have called Jim an "apple-polisher," "teacher's pet," or something much more impolite and probably obscene. In college, we call Jim a "survivor." Now that I've thoroughly embarrassed you, Jim, what is a protein, please?

JIM: It is a macromolecule made of chains of amino acids.

FARNSWORTH: Precisely, and now, somebody else, what is an enzyme?

ANOTHER STUDENT: An *enzyme* is a compound that speeds up the rate of a reaction.

FARNSWORTH: Wonderful. That is absolutely correct, but in biological systems, the effect of the presence of an enzyme, for all practical purposes, means that a particular reaction will go. Without the enzyme, at biological temperatures, the reaction would simply not go far enough to be significant. What this signifies in practice is that if you have a couple of reactants, and you want a particular product, those reactants will just sit there forever until you add the appropriate enzyme. Then, whiz, the reaction goes to completion. So functionally, an enzyme acts as a trigger to make a reaction proceed to completion in a biologically reasonable time.

Since almost everything a cell does involves specific chemical reactions performed at specific times, it stands to reason that by controlling which enzymes are produced when, you can control everything the cell does. That is why the production of enzymes is so important; critical, really. What we need, then, is some kind of mechanism that will allow cells which have divided to manufacture enzymes the same way the parent cell did.

There is an everyday situation that will give you some idea of how this works. Suppose I am a contractor and I am building a set of houses on a subdivision tract. The basic design of the houses will be identical—same kind of furnace, same kind of plumbing, same kind of floor plan—but I'll allow the crew chief in charge of each house to have a little discretion about minor things—placement of the windows, paint scheme, and so on—but functionally, they'll be the same.

I now hire an architect to draw up a set of plans. She's a very expensive architect, and it would cost a fortune to have a new set of plans drawn if something happened

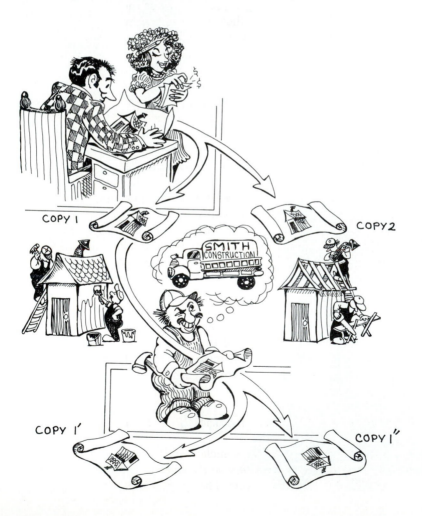

to the originals, so do I send the originals to each of the house sites, let a house be built from the originals, and send them on to the next job site? Of course not—I make an exact *copy* of the plans to send to each house site.

Of course, if I do that, there is nothing to prevent one of the carpenters on the job site from deciding he wants to be a contractor too and taking his copy of the plans, copying them, and starting the process all over again. I think you can see that this process could go on indefinitely, so long as a true copy can be made of the plans each time. What we have here, then, is a copying mechanism that duplicates an original set of instructions for doing something, in this case, making a house; the copy, in turn, can then be copied. Each new copy becomes the "original" for the next copy, generation after generation.

The same general idea operates in the cell. You have a set of plans, or instructions, for making protein enzymes. That plan can be duplicated so that each divided cell will have a set of the instructions. When this generation of cells divides, the instructions will be copied again, then again, for as long as the line of cells lives.

People looked for a long time before they found out where in the cell these instructions were located. Just as the contractor would keep the original plans in a safe place, the cell keeps *its* plans in the safest, most centrally located and protected place it has—the nucleus.

After an enormous amount of effort, it was discovered that there is a certain kind of molecule found in the nucleus that has exactly the properties we are looking for. One, it can be copied, and, two, it can store a tremendous amount of information which can be used to specify how particular proteins should be made. This molecule, which is a macromolecule, is called **deoxyribonucleic acid**, or, more familiarly, **DNA**. DNA belongs to a class of compounds called **nucleic acids**, which we will examine in some detail.

DNA is called an *informational macromolecule* because its importance is not in what it does, as with a hormone, but in the information stored in its structure. It is found in the nucleus, as I said, but it is also found in mitochondria and in chloroplasts.

Like most macromolecules, it is of variable length and is made of repeating subunits. Just checking now; what were the repeating subunits of a protein?

STUDENT: Amino acids.

FARNSWORTH: Right. Well, the subunits of DNA are called **nucleotides**. Very important not to get these mixed up. Proteins are made of amino acids. DNA is made of nucleotides. There are two classes of nucleotide used in DNA, and two individual kinds of nucleotide in each class, making a total of four nucleotides. How many kinds of biologically important amino acids are there?

STUDENT: Twenty.

FARNSWORTH: Good. Twenty different amino acids, four different DNA nucleotides.

Let's look at a generalized structure for a nucleotide; then we'll look at each of the individual ones. (*He places a transparency on the overhead projector.*)

There are three components to a nucleotide. The first is a 5-carbon sugar. You're

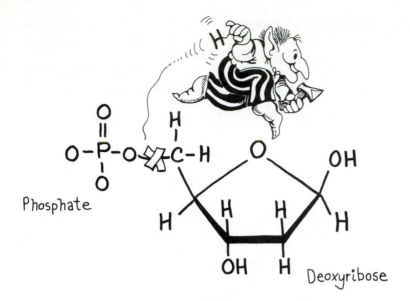

Phosphate

Deoxyribose

bons in the ring and another in a side group. The 5-carbon sugar in DNA is called *deoxyribose*—hence the name *deoxyribose nucleic acid*. There is another 5-carbon sugar, ribose, that's found in the other nucleic acid, RNA, or ribonucleic acid, but we'll come to that later.

We are going to fasten an old friend to our deoxyribose. Remember phosphate from ATP? Well, ol' phosphate really gets around, because, as you can see in the figure, we knock a hydrogen off the deoxyribose side chain and bond a phosphate to it.

So far, so good, but now it starts to get just a little bit complicated. To complete our nucleotide, we are going to bond on to the sugar something called a *nitrogenous base*. (*He begins drawing the bases on a transparency on the projector.*) There are two kinds of nitrogenous bases. One type has a single-ring structure and is called a *pyrimidine*. The other kind has a double ring and is called a *purine*. In DNA, there are two different pyrimidines. One is called *cytosine*, and the other is called *thymine*. As you can see, they are similar, but the side chains are different and are in different positions on the rings.

There are two kinds of purines, also—one called *adenine*, the other *guanine*. Both are double-ring molecules, both are similar, but like the pyrimidines, there is a difference in the nature and location of the side chains.

I have to pause here to give you fair warning about something. It is *extremely important* that you remember these names and remember which is a purine and which is a pyrimidine. I also want to tell you about a potential trap. One of the purines is called *adenine*, right? Well, there's an amino acid you'll frequently see mentioned in your reading called *alanine*. Remember it by: adenine is in *DNA*. Okay, sing out your answers—gimme a pyrimidine that starts with a *T*.

THE WHOLE CLASS: Thymine!

NITROGENOUS BASES

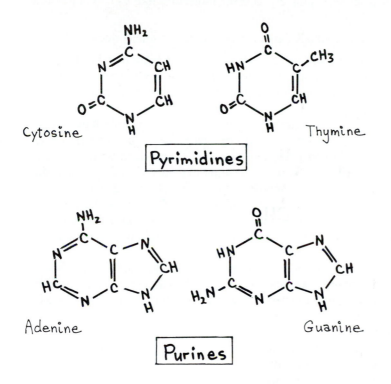

Cytosine   Thymine

Pyrimidines

Adenine   Guanine

Purines

FARNSWORTH: Gimme a purine that starts with a *G*.

THE CLASS: Guanine!

FARNSWORTH: Gimme a pyrimidine that starts with an *R*.

Confusion.

FARNSWORTH: Good. There isn't one. Now, finally, we are ready to build a nucleotide. We had our deoxyribose sugar bonded to a phosphate. Now we bond a nitrogenous base to one of the carbons on the deoxyribose ring, and lo and behold, we now have a nucleotide. Actually, we have one of four different kinds of nucleotides—they're named after the organic base that is used, hence, adenine nucleotide, thymine nucleotide, and so on. Okay, question time. Any questions?

STUDENT: Yeah, how do you get nucleotides?

FARNSWORTH: How do you *get* nucleotides?

STUDENT: Uh, yeah, like, where do they come from? Do you get them by eating them, or what?

FARNSWORTH: Oh, I see. Well, yes, animals can acquire a supply by eating cells, either animal or plant, which already contain DNA. The DNA is broken down into nucleotides, which the animal can then reassemble to make its own DNA. Plus, as you can see, the nucleotides are not very complex molecules, and plants can make them from scratch from very simple precursor molecules like CO_2. Okay? More questions?

STUDENT: Do different animals have different kinds of nucleotides?

FARNSWORTH: Ah, good question. No, they all use the same four kinds in their DNA. What differs is the proportion of each kind, and the way they are arranged on the DNA molecule. That question provides a good lead-in to our next step, the construction of a DNA molecule.

The first thing to do is make a chain of nucleotides. We take a couple of nucleotides, arrange them so the nitrogenous bases are on the same side of the deoxyribose, and bond one of the oxygens on the phosphate of one of the nucleotides to one of the carbons in the deoxyribose ring of the other. A couple of points to consider. First, since we are building a polymer, we could bond as many nucleotides together as we wanted or needed. We want 100,000 nucleotides strung together? No problem. Second, there is nothing in the structure of the nucleotides to prevent us from bonding any nucleotide side by side to any other one. Purine next to purine *or* pyrimidine. Adenine next to adenine or any other kind of nucleotide.

What we have now is a single, potentially very long, strand of nucleotides bonded together side by side. Now comes the neat part. DNA is not a single strand of nucleotides—it's a *double* strand. There is a second strand of nucleotides alongside the first.

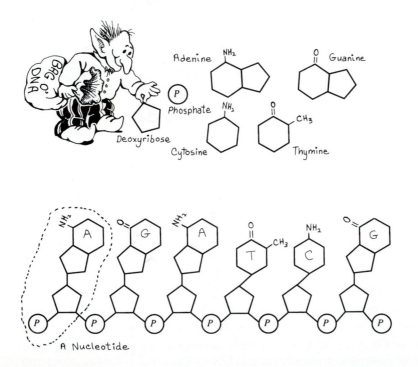

We *could* make two single strands completely independently, lay 'em next to each other, somehow try to bond them together by the nitrogen bases, and call the whole mess DNA, but it doesn't work that way, and now I'm going to show you how it *does* work.

First, we have two strands of deoxyribose and phosphate, and we're going to set them down like the side rails of a ladder, the same distance between the rails for the whole length of the ladder. Now, on one of the strands, we're going to attach our nitrogenous bases. For right now, it doesn't matter which ones we pick. Let's say adenine-guanine-adenine-thymine-cytosine-guanine.

Okay, one strand is complete. Now we have to plug in the nitrogenous bases on the other strand, but before we do, we have two rules that we have to observe. First, when we put a nitrogenous base in opposite one that is already in place on the other side, we can't put in a base that is so large it will tend to push the two sugar-phosphate strands apart. Second, it can't be so small that it will leave a big gap between the two opposing nitrogenous bases. If you had a gap, the two sides could not be joined together with the special kind of bond I'll talk about in a minute or so. It has to fit just so. Okay, what was our first base on the original side? Adenine, right? Adenine is a double-ring purine, and, as you can see, if you try to stick another purine, either adenine or guanine, in on the other side, it won't fit. We could try cytosine, but if we looked closely, we would see some small gaps. The only thing that will fit is a thymine. The next base down the line is guanine. Also a

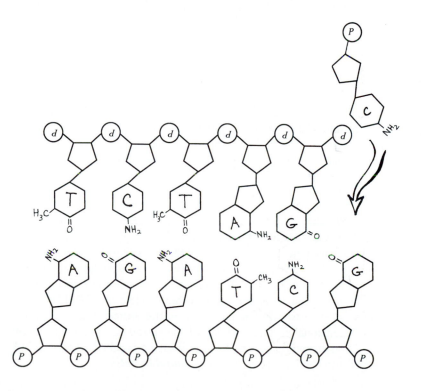

double-ring purine, so we can't use another purine on the other side, and if we try thymine, we'll get small gaps. The only thing we can use is cytosine. The same is true for the next guanine. Looking at the next base, thymine, which has one ring, we see that if we tried to put another thymine or a cytosine on the opposite side, we'd have a big gap. If we used guanine, there'd be some small gaps. Now, before we go any further, can anybody see any generalizable rules about what you have to use on the other strand, if you have a given base on the first strand? Think for a second. Yes, Caro—no, Cheryl.

CHERYL: A purine can bond only with a pyrimidine, and a pyrimidine can bond only with a purine.

FARNSWORTH: Terrific! Now, does anybody want to make that even more specific?

ANOTHER STUDENT: Adenine can bond only to thymine, and vice versa, and guanine can bond only to cytosine, and vice versa.

FARNSWORTH: Wow! I'm impressed. I thought I'd have to go a couple more down the chain. Absolutely correct. If both sides of the molecule are going to stay parallel and there aren't going to be any big gaps between opposite bases, A has to bond to T, and G has to bond to C. Purines are thus *complementary* to pyrimidines. Adenine is complementary to thymine, and guanine is complementary to cytosine.

Now you are all ready for an interesting question. Suppose I gave you one strand of a DNA molecule plus a box of assorted nucleotides, and I said to you, "You have never seen the original double-stranded DNA molecule this strand is from. Using only the single strand I just gave you and the nucleotides in the box, I want you to recreate the original molecule." How much in the way of brains would it take to perform this task?

STUDENT: Not much.

FARNSWORTH: That's right! You could even be a (*Here he mentions the name of his school's hated rival*) student and do it. Why is the job so easy? Because of complementarity. If you started at one end of the strand you were given, and let's say it was an adenine nucleotide, for the first nucleotide on the other side, you would reach in the box and pull out a...?

STUDENT: Thymine.

FARNSWORTH: —nucleotide. You'd then go down, tick, tick, tick, and put a complementary nucleotide at each nucleotide location.

All right, let's summarize where we are so far. DNA is a double-stranded molecule made of two chains of complementary nucleotides. It is a very large molecule, primarily found in the nucleus of the cell. We have to know one more thing about DNA; then we can go on to the structure of the other nucleic acid, RNA.

So far, I've shown DNA in two dimensions, flat on the projector screen. DNA actually has a three-dimensional structure. Each of the strands is coiled, like a coil spring or a "Slinky." Something that looks like a coil spring is called a *helix*. DNA has two strands coiled into helices, so it is called a *double helix*.

There are a number of ways you could arrange two coils wrapped around each other. (*He puts a picture of several types of helices on the projector.*) One way kind of looks like a corkscrew. That isn't the way DNA works, though. With DNA, imagine you had the cardboard tube from the middle of a roll of paper towels, and you spiraled two parallel strands of spaghetti, cooked, of course, down the length of the tube. Wait 'til the spaghetti hardens, then throw away the tube. The two spaghettis represent the sugar-phosphate backbones of the molecule. The bases meet in the middle of the cylindrical space between the spaghetti strands.

One thing you should know about the way the bases are fastened together is that they are tacked together with a special kind of bond called a ***hydrogen bond***. That's *bond*, not *bomb*. A hydrogen bond is called a *weak bond* because it doesn't take much energy to break it. What this means is that the two strands of the DNA molecule are fairly easy to separate from each other. It is much easier to separate the two strands than it is to break the molecule apart crossways between a phosphate and a sugar. This observation will be very important to us later.

All right, I think that's enough about DNA structure for now. I just want to talk about the structure of ***ribonucleic acid (RNA)***, and we can call it a day.

RNA is a single-stranded nucleic acid, but it is not found only in the nucleus. It moves all over the cell, as we will see in a lecture or two. The nucleotides making up RNA have ***ribose***, instead of the deoxyribose found in DNA. This means there

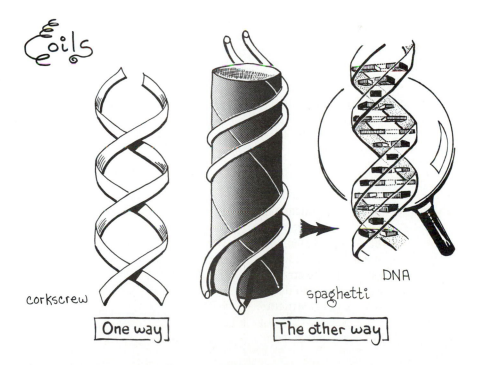

Coils

corkscrew

spaghetti

DNA

One way

The other way

are two sets of nucleotides available in the cell—ribose nucleotides for making RNA, and deoxyribose nucleotides for DNA. There's one final difference between DNA and RNA. Three of the four ribose nucleotides have the same nitrogenous base used for the DNA nucleotides—adenine, guanine, and cytosine. Where DNA would have thymine, though, RNA has *uracil*, another pyrimidine. Uracil is complementary to adenine, just as thymine is.

Okay, what did we do today? We learned about the structure of the two types of nucleic acids, DNA and RNA. Both are polynucleotides, made up of chains of nucleotides, which are in turn made of a 5-carbon sugar, a phosphate, and a nitrogenous base. We learned that DNA is double-stranded and that the base pairs on opposite strands are complementary—purine with pyrimidine, A with T, G with C. RNA is single-stranded, has ribose instead of deoxyribose, and substitutes uracil for thymine.

Wait! Before you go, let me remind you that we have had seven lectures so far, some of them industrial strength, and before you know it we will have—an exam. Whatever you have to do at this point—dump your friends, quit your job, move out of your house—DON'T FALL BEHIND! It will be very, very difficult to catch up. I won't wait for you. Also, capitalize on the help resources. Come see me during office hours, and come to the review sessions. If you find yourself in deep dung, and I find out that you could have gone to a review session but didn't, do you know how much sympathy I'm going to have for you? An amount so small it couldn't be detected by the finest electron microscope in the world. We have only a few more molecular lectures; then we can relax a bit. Okay, I'll see you next time. Now I have to catch a little sunlight. (*His burnoose swirling around him, he exits.*)

Day 8

DNA Replication

Professor Farnsworth enters, wearing a Harris tweed jacket.

FARNSWORTH: Good morning. Good news for you today. I'm going to show you only one new thing, and it's really not that complicated, but we'll look at it a couple of ways, and then maybe we'll have a little review toward the end of the session. First, though, what did I talk about last time?

STUDENT: DNA and RNA.

FARNSWORTH: Fine, what about DNA and RNA?

SAME STUDENT: DNA is a double-stranded polymer made of nucleotides, which are made of deoxyribose, phosphate, and one of four nitrogen bases.

FARNSWORTH: Good. Yes, DNA is double-stranded, and that observation is important in what we are going to look at today—how DNA copies, or replicates, itself.

Let me set the problem for you. You have this DNA in the nucleus of the cell, and now it comes time for the cell to divide. You couldn't just take a knife and cut the cell in half, cutting the DNA molecule in half in the process, because each new cell would have only half the original DNA. Wouldn't work. We have to do something a little fancier.

Think about copying for a second. Suppose I have a magazine article I want copied. I take it down to the copy machine, push the button, and it spits out the copy. Now, have I altered the original page *in any way?* No, my original is absolutely intact, and the copy is a whole new thing, which doesn't have any physical part of the original document—ink, paper—in it. The only thing in common between the copy and the original is the information. This would be called *conservative replication* because the original is whole, or conserved. DNA replication doesn't work this way. There is no cellular copying machine that would let you do this. Instead, DNA replication works by making use of a property of the DNA molecule we looked at last time. Remember *complementarity?* What was that, remind me.

STUDENT: With complementarity each base on one strand of the molecule can match up with only a particular kind of base on the other side. I think it was guanine matched with cytosine, and thymine matched with alanine.

FARNSWORTH: *AAACCK!* Okay, okay, that's good, that's good, you got it basically right, but it's *adenine*, not alanine. Alanine is the amino acid. Don't feel bad; it took me two years before I got it straight. Sure, remember we noted that if somebody gave you one strand of DNA, you could make the other strand if you had a box of DNA nucleotides? That is in effect what you do. You take your double-stranded DNA, split it into two strands, and assemble a complementary strand on each of the separated strands. Each of the separated original strands acts like a *template*, or pattern, for the new strand. When you're all through, you have two double-stranded DNAs, each of which consists of one of the original strands and a complementary copy. That's why this kind of replication is called *semiconservative*, because each offspring DNA has half the original molecule. Now let me ask

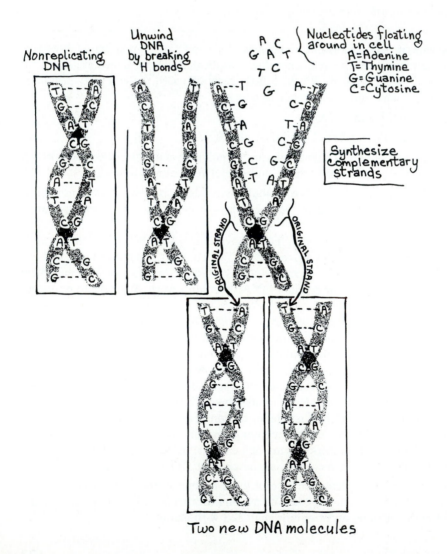

Two new DNA molecules

you something. Let's say that you had an original cell, with its original DNA. The cell now divides, and the DNA divides semiconservatively, and now you have two offspring cells. First interesting question. Where is the original cell now?

Animated discussion and confusion.

FARNSWORTH: Ha! Well, when you figure it out, let me know. It probably doesn't exist anymore, but don't ask me where it went. I love things like that. Here's one I gave to last semester's class. Suppose I gave you a multiple-choice question that looked like this (*writes on projector*):

A dog is:
   A. A cat.
   B. Blue.
   C. Bigger than an elephant.
   D. Fueled with gasoline.
   E. There is no correct answer to this question printed on this exam.

Now, what's the answer? (*Puzzled looks and conversation*) Well, this is something that is called a *self-contradictory statement,* and they're one of the most fascinating parts of language. Maybe I'll give you more later, but let's get back to the DNA, and my second question. You have these two cells, right, each with half the original DNA strand. Let's say each of these cells divides again, so now you have four cells. Is it possible that any of those cells will have *both* the original strands? Wait—before you answer—is it possible that any of those cells will have *neither* of the original strands?

STUDENT: Yes.

FARNSWORTH: Yes, what?

THE STUDENT: Uh, yes, one of the cells could have both the original strands.

FARNSWORTH: *Wellll,* think carefully about that for a second. In your first generation of divided cells, each DNA double helix in each of the two cells consists of one original strand and one synthesized strand. Now, when each cell divides (the second division), the original strand goes to one cell; the synthesized strand goes to the second cell. The original strand will have synthesized a complementary strand. The synthesized strand will also have synthesized a complementary strand. Now you have one cell that has an original strand and a synthesized strand, and another cell that has two synthesized strands. So, in the second generation, you have four cells, two of which have an original strand and a synthesized strand, and two of which have two synthesized strands. In the *third* dividing generation, how many cells will you have?

STUDENT: Eight.

FARNSWORTH: Good, and what fraction of those will have an original strand?

SAME STUDENT: A quarter of them.

FARNSWORTH: Good, now will that fraction of each cell generation that still has an original strand get bigger, smaller, or stay the same with each successive division?

A jumble of answers pour forth from around the room. "Bigger." "Smaller." "It'll stay the same."

FARNSWORTH: (*Amused.*) Well, all right, you're thinking, at least. It will get smaller with each generation because you still have only two cells that have an original strand, but the total number of cells in the generation might be in the millions. Now, what does this tell you about how accurate the replication process has to be?

STUDENT: It has to be pretty accurate.

FARNSWORTH: Right. What you are essentially doing is making a copy of a copy, for many generations of copies. If you want to spend about two or three bucks, and spend 10 minutes, you can do a very illuminating experiment. Get a magazine article or any kind of printed document you want copied. Go to a copy machine and make a copy. Now, feed the copy, not the original, into the machine and make a copy of the copy. Then make a copy of the copy of the copy. Just keep doing this, and write the number of each generation of copy on each of the copies. After the fifth generation or so, depending on how good the machine is, you will see mistakes—distortions, marks on the paper, and so forth. By the twentieth copy, you won't be able to read some parts, and in most cases, by the time you get to the fiftieth generation, you'll have total garbage. There are theoretical reasons why it is impossible

to have a perfect copy of *anything* and errors tend to compound with each generation of copying. When applied to cell division, where you might have a hundred cell generations before the organism dies, what does this tell you about the copying mechanisms?

STUDENT: There has to be some mechanism for correcting errors during each generation of division.

FARNSWORTH: Exactly. Adenine is supposed to be a complement only to thymine, but what happens if accidentally you get a guanine instead? There has to be some way of, first, determining that an error has been made and, second, correcting the error.

In addition to guarding against errors in the copy process, the DNA has to be able to repair damage done to it between replications. It's a tough world out there, gang. You go out in the sun, and the ultraviolet rays in sunlight can knock a nucleotide out of place in the DNA in your skin cells. You sit on the right kind of granite rock, and your gonads' DNA gets zapped by atomic radiation. You suck in a nice, big, rich lungful of cigarette smoke, and the DNA in the cells lining your lung gets bathed in chemicals that can scramble the order of the nucleotides. Ain't fair, but there it is. DNA is not helpless against this, however. For example, there are a number of repair mechanisms that let the DNA replace a nucleotide that has been knocked out of position by ultraviolet light. I'll come back to this and explain how some of these correcting mechanisms work, but first I want to expla—, hold it, there's a question. Yes?

STUDENT: (*She has a slightly hostile, vaguely suspicious attitude.*) You've been talking about this DNA's repairing itself, and being able to correct mistakes, as though it *knows* how things should be, as if it's thinking about it. I mean, is there some little man in the DNA that decides that it's been injured and then fixes it? How does the DNA know that it has been injured, and how does it know how to fix itself? Is it alive, like the motorcycle?

FARNSWORTH: (*Momentarily taken aback*) Wow! What a great question. Sure, it certainly does look like this molecule can react to stimuli, and show different behavior depending on what the stimulus is. (*Pauses, thinking*) I think I can answer your question; my problem is figuring out how to explain it without getting too philosophical or complicated. Let's try a crude, rough answer first and see if it makes sense.

Gold. The metal gold. What color is it?

THE SAME STUDENT: Kind of yellow.

FARNSWORTH: What purpose does the yellow color of gold have, in terms of benefiting the gold?

THE STUDENT: None.

FARNSWORTH: Then why is it yellow? Why isn't it brown or green?

There is much discussion, and finally a student in the front row speaks up.

Ways to injure DNA

THE STUDENT: I'll try an answer. There isn't any reason for the color, no functional reason; it's just a property of the metal that it reflects yellow light. Something about the gold molecule reflects yellow light.

FARNSWORTH: Okay, file that answer away. Now consider a jeweler, a jeweler who experiments with different kinds of metals to make inexpensive jewelry. He's discovered a wonderful alloy metal. Cheap. Abundant. Easy to work. Unfortunately, it's kind of a greenish-brown color. But the jeweler thinks to himself, "If I just added a little bit of this pigment, I could change it to yellow." Okay, now, does the yellow color of the alloy have a purpose or function now? Yes, to make it look like gold. Did the yellow alloy get to be yellow through the action of a designer; did it get to be yellow as the result of some process of thought? Yes.

No Purpose   Purpose

So now we have two pieces of metal—one gold, one alloy. To us, they both look alike; identical in fact. But one is yellow as the result of the natural property of the material; the other is yellow because of the conscious action of a designer. To *look* at the two, you couldn't tell which is which. Now, we go back to DNA. It *could* be that the capacity for self-repair is the result of a little tiny dwarf who lives in the molecule and knows what to do, but it is more likely that the capacity for self-repair is the result of a *natural property of the molecule itself*. So the answer to your question is that the DNA molecule doesn't *know* how to repair itself; the ability to repair, just like the ability to replicate, is a result of the peculiar structure of the molecule. You didn't ask it, but I'll bet you there is a second part to your question, and that is, "It is absolutely remarkable that this molecule can do all these things; how come this is the only molecule that can do all this?" That, too, is a wonderful question, but I will have to defer the answer until we talk about something called *natural selection* later in the semester. Don't forget the question, though.

Now let's look at the replication process in a little more detail. We start with a double strand of DNA. The first thing you have to do is break the hydrogen bonds that hold the two strands together. You do this with a class of enzymes called **helicases**. Once the hydrogen bonds are broken, the strands naturally tend to spring apart, and to keep them apart, you insert something called a **single-strand binding protein**.

Once the strands are separated, you are ready to start replication. Your first step is to tack a short complementary length of *RNA* at the end of the exposed strands. This RNA is called a **primer**, and without it you don't have replication.

What we have now is a thing that looks like a letter *Y*. The point at which the two strands unwind is called the ***replication fork***. To look at it, you might say, "Okay, I'll start adding nucleotides up here at the end of each arm of the Y and work my way down to the fork." Except it doesn't work that way, and the reason it doesn't is because each individual strand of DNA has a head and a tail, like an arrow. The head end is called the 5′ end, which is pronounced "five prime," and the tail is called the 3′ end. The two strands lie next to each other head to tail, so the 5′ end of one is next to the 3′ end of the other. When the two strands unwind, the exposed 3′ end does just what you would suppose it might, and starts adding nucleotides one at a time, working toward the replication fork. The new strand is thus synthesized in the 5′ to 3′ direction. The other strand, with the 5′ end, is going to give us problems though.

It turns out that DNA can be synthesized *only* in the 5′ to 3′ direction. So we *can't* start at the exposed 5′ end and work toward the fork. Instead, on this strand,

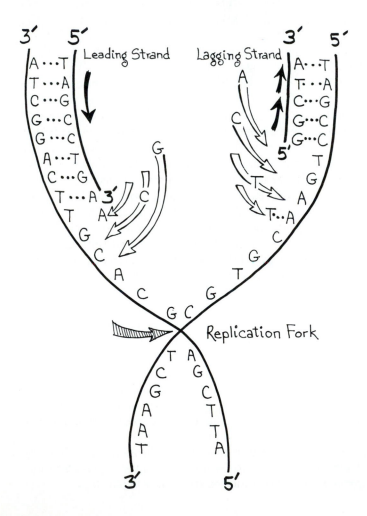

what we have to do is unwind the original strand a little way, start at the fork, and synthesize back toward the end. We then unwind a little more, start at the new location of the fork, and work backward toward the 3′ end of the length of strand we just made. We use an enzyme called **DNA ligase** to join the 3′ end of the strand we just made to the 5′ end of the strand made just before it. We unwind some more and make a new length of strand, again starting at the fork and going backward. The strand that's made on the original 3′ strand is made continuously, working toward the constantly moving fork. This new strand is called the **leading strand**. The strand that's made on the original 5′ strand isn't made continuously—it's made in chunks, starting at the fork and working back up toward the 5′ end. This new strand is called the **lagging strand**. The net result is that the new strands are always made in the 5′ to 3′ direction. Now, I don't know about you, but it just blows me away how clever the people were who figured this out.

The enzyme that links the nucleotides together is called **DNA polymerase**, and it not only serves that function but also performs proofreading duties. In a way that is not fully understood, if the wrong nucleotide is put into position in the newly synthesized strand, the DNA polymerase somehow "knows" that it is the wrong one, removes it, and puts in a correct one.

In addition to proofing replication, enzymes can help in repair of damage to a nonreplicating DNA molecule. Let's say radiation, or UV light, or whatever knocks one nucleotide out of place and shoves in an incorrect one. First question: how do you know something is wrong? Well, remember, when we were talking about DNA structure, I said that you wanted to have the two sides parallel. If you had the wrong nucleotide, there would be either a bulge or a pit in the molecule—the two sides wouldn't be parallel anymore. In effect, what a repair enzyme does is *look* down the molecule for such a pit or bulge and then proceed to cut out the damaged section and replace it with the correct nucleotides.

I hope you realize that when I say this enzyme is "looking" for a pit, I don't mean the molecule has eyes or belongs to Local 235 of the International Brotherhood of DNA Repairers. When I say "looking," that is a fast way of saying "the repair enzyme has the molecular property of binding to an area of the DNA molecule which has been distorted by the presence of an incorrect nucleotide." To give a human quality to something that is nonhuman is called *anthropomorphizing*, and it *is* a quick way of giving you a rough idea of something; however, it can be dangerous if you don't realize that it is just a shortcut way of saying something, and isn't literally true. Okay, any questions?

STUDENT: How come you have two strands of DNA? Couldn't you have a self-replicating molecule that has only one strand?

FARNSWORTH: Terrific question! Now I have a terrible answer for you. Nobody really knows. Yes, you could have a system of self-replication that involves only one strand. You would first make a complementary strand off your original strand, using the original strand as a template. You would then make a complementary strand off the one you just made. This second strand would be a complement of a

complement, or exactly like the original. So you could make a duplicate of your original strand, for sure, but do you see any problems with this?

STUDENT: It's a two-step process; you're making a copy of a copy. That doubles the chances of making a mistake.

FARNSWORTH: Okay, any other problems?

ANOTHER STUDENT: If this is going to be for a dividing cell, you need two copies of the original, one for each of the offspring. This way, you'd produce only one copy.

FARNSWORTH: True, but remember, you still have the original, unlike the double-strand situation, where the original doesn't exist anymore. The original could go to one offspring; the copy could go to the other. However, the two-step process, and the possibility of error, argues strongly for using two strands. In addition, in the two-strand system, you can always use the second strand to proof the newly synthesized strand. But to tell the truth, these are just ideas; nobody knows for sure. All right, everybody put your notebooks away, and get a pen. C'mon, clear those desks off. We're going to have a quiz. (*Groans, complaints from the class.*)

As the guy said as they strapped him in the electric chair, "Well, this has certainly been a good lesson for me. I'm never going to do *that* again." I've been telling you how important it is to study as you go and not wait until the night before the exam. If you've been studying regularly, you won't have any problems with this— they're super-easy questions. If you haven't been following along, you'll get your nose rubbed in it. Okay, the assistants will pass out the quiz. Read the instructions; you have five minutes. Go.*

Editor's note: The reader may wish to try Professor Farnsworth's quiz, which follows.

**Biology 100
Review Quiz**

There is no virtue like necessity.
—William Shakespeare
Richard III

Name (print) _____
 Last   First   Initial

Student Number _____

Instructions: The correct answer is true in all its parts, and relevant to the question. There is only one correct answer to each question. Do the easy ones first, then the tough ones. Good luck. You'll need it.

1. *Life* is:
    A. A quality held by all things that can respond to stimuli.
    B. A magazine.
    C. Something that stops when the brain stops functioning.
    D. Defined by the power of reproduction.
    E. Found only in the presence of oxygen.

2. Which of the following is the best example of a *mu* question?
    A. How much is two plus two?
    B. In which year did Abraham Lincoln sign the Constitution?

    C. What color is your car?

    D. When did World War II end?

    E. Assume that you have a supply of nucleotides. If you strung them together into a polymer, could you have more than four nucleotides in a row?

3. To attach two glucose molecules together to form a disaccharide, you would need to use the process of:

    A. Dehydration.

    B. Electron transport.

    C. Hydrolysis.

    D. Hydrification.

    E. Malting.

4. One would find a peptide bond in:

    A. A polysaccharide.

    B. A fat.

    C. A glycolipid.

    D. A protein.

    E. RNA.

5. In cell energetics, the significance of NADH is:

    A. That it can be broken down into ATP.

    B. That it can be combined with NAD^+ to form pyruvate.

    C. That it can carry electrons and protons.

    D. That it is one of the first by-products you get after you split the glucose.

    E. That it has two, not one, high-energy bonds.

6. What do you use ribulose biphosphate for?

    A. To combine with CO_2, in the first step of making glucose out of CO_2.

    B. To pick up light energy to drive the Calvin-Benson cycle.

    C. To oxidize NAD^+ in the Krebs cycle.

    D. To pick up photons in the dark cycle.

    E. To prevent the fluorescence of chlorophyll.

FARNSWORTH: Okay, that's enough time. Now I'm going to tell you the answers, and you can score them yourselves. And just to show you I'm in a good, mellow mood today, you don't have to turn them in. I'm serious about keeping up though. You don't keep up to be ready for pop quizzes; you keep up to be sure that you don't fall behind and have to do panic studying before an exam. Okay, what's the answer to the first one?

STUDENT: *B*. A magazine.

FARNSWORTH: Sure. Remember the conditions. The correct answer is true in all its parts, and relevant to the question. *A* was not right, because machines respond to stimuli and we can't agree that machines are alive. Same problem with *C* and *D*—there are exceptions to both cases. *E*'s no good, because you now know there is anaerobic respiration. But actually, you wouldn't need to know any biology to answer the question. Look at the word *Life*. It was italicized. This is typographically the way proper nouns such as books and magazines are treated. Of all the answers, only *B* suggested the need for such typographical handling. Tricky, but you have to read carefully.

How about question 2 now? A *mu* question contains an inherent assumption which may not be true. Remember, the question asked which was the *best* example. The question "How much is two plus two?" contains an assumption, but it is the assumption that there is an answer to the question, and this is something true of most questions. On the other hand the Lincoln question carried a clear false assumption—that he signed the Constitution—so it provided the best answer. I hope nobody fell for the trap provided by *E*, which was clearly the longest answer.

The third question was straightforward. You knew it or you didn't, and if you knew it, the answer was "dehydration."

Question 4 had an interesting twist. An unfamiliar word, *glycolipid*. Now, when you confronted that word, you had to ask yourself "Have I just forgotten this word, or is it a word that I haven't seen before, and maybe I'm supposed to figure something out about it?" Well, here's where you have to do a little educated guessing. If you *knew* that a peptide bond was in a protein, which it is, you could disregard everything else. If you *didn't* remember that, you would have to exercise a little logic and play some odds. Here is a word that you maybe haven't seen, *glycolipid*, but you *have* seen the parts. *Glyco-* in *glycolysis*, which has something to do with sugar, and *lipid*, which has something to do with fat. With that knowledge, you could estimate it as a possibility.

Question 5 is a little tougher. You have to know not what NADH *is*, but what it *does*, and the question is not just a statement from your lecture notes turned around into a question. If you haven't studied at *all*, you can't even start to make an educated guess, but you can decrease the likelihood of some of the alternatives by doing a little thinking.

For example, look at *A*, "it can be broken down into ATP." Well, *nothing* gets directly broken down into ATP. ATP isn't a monomer, like an amino acid, which you can use to build a larger molecule. So you could probably eliminate that one. Consider the last one, "That it has two, not one, high-energy bonds." Well, you probably remember that it is ATP and ADP that have high-energy bonds, so you could dump that answer. That leaves three alternatives, which is a lot, but still a lot better than five, and you might get lucky and pick *C*, the electron carrier answer.

Question 6 is probably the pickiest question of the lot. Ribulose biphosphate is very important, to be sure, but you learned about it along with a whole pile of other new words. The only real way you could know the correct answer, *A*, is through careful study.

Okay, that's about it for today. One more lecture about DNA next time, then we can relax for a bit.

Day 9

Transcription and Translation

Professor Farnsworth enters, wearing slacks and a Norwegian fisherman's sweater. As the semester progresses, his dress becomes progressively more informal.

FARNSWORTH: Good morning. One more day of molecules, just as I said, then we'll start talking about things you can see. First, what did we talk about last time?

STUDENT: DNA replication.

FARNSWORTH: Good. Details please.

SAME STUDENT: Well, the DNA unwinds, then you make a new strand off of each of the old ones, but on one of the strands, you start working toward the replication fork from the end of the strand, and on the other strand, you work from the fork back to the end.

FARNSWORTH: Nicely stated. Keep in mind the fact that the DNA unwinds because that will be very important for what we'll talk about next.

Okay, remember a couple of days ago, I made an analogy about a contractor making plans and then shipping a copy of the plans to the job site so the workers could work from the copy and not risk damaging the original? Well, today I'm going to tell you how the molecular DNA contractor works, but this contractor builds protein, not houses.

DNA is a double-stranded molecule, and we're going to make a copy of the information on the DNA molecule that will tell us how to make a particular protein, but we don't have to copy the whole molecule, as we did in DNA replication. What we do is unwind a section of the DNA molecule that has the information we need— it doesn't have to be at the end of the DNA; it could be in the middle, like this. (*He puts a transparency on the overhead projector and draws a picture of a stretch of DNA.*) We have the two strands separated now, which means we can get at those nitrogenous bases.

Next, we pick one strand or the other. It seems that for some proteins you pick one side, and for others you pick the other side, but the point is that you use only one strand here—the other doesn't play a direct role. Once you've picked this one strand, you proceed to synthesize a complementary strand off it, just as you did when you replicated DNA. There are a couple of differences in the nucleotides you use, though.

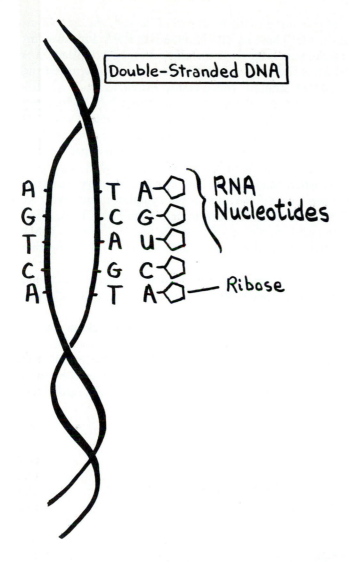

In DNA synthesis, you used DNA nucleotides as your raw materials, but here we will use different nucleotides, called **RNA nucleotides**. An RNA nucleotide uses ribose instead of deoxyribose, and there isn't a thymine RNA nucleotide; instead you have a **uracil** RNA nucleotide, which fits where thymine would.

So you go down your selected strand of DNA, and you do just as you did with the leading strand of DNA replication—you add complementary nucleotides, until you get a strand of—what?

STUDENT: RNA.

FARNSWORTH: Good. RNA. You then detach the RNA strand from the strand of DNA and let it break free. The DNA then winds itself up again. The strand of

RNA leaves the nucleus and goes out into the main part of the cell, the cytoplasm, where the protein is actually made. There are different kinds of RNA, but since this kind of RNA essentially carries a message from the nucleus to the cytoplasm, it is called *messenger RNA*, which is abbreviated *mRNA*. The process of synthesizing the messenger RNA on the DNA strand is called *transcription*.

While transcription is taking place in the nucleus, something is happening out in the cytoplasm. Floating around in the cytoplasm are millions of amino acid molecules, all 20 different kinds. Remember, we want to make proteins out of these amino acids, and a protein is just a bunch of amino acids arranged in a specific way. The information contained in the sequence of nucleotides on the DNA is going to tell us which amino acid goes where, and now we have a complementary transcript of the DNA molecule, the messenger RNA, out in the cytoplasm. However, we can't use the messenger RNA directly to say how to tie the amino acids together. We have to attach the amino acids to a kind of molecule that can connect to the messenger RNA in a particular way, and then we can tie our amino acids together to make a protein.

This connecting molecule is called *transfer RNA*, or *tRNA*, because it picks up, or *transfers*, amino acids from where they're floating around to the area where the proteins are made.

Transfer RNA is about 80 nucleotides long, and it is all folded back on itself so that it looks something like a cloverleaf. About in the middle of this molecule is a critically important sequence of three nucleotides, which form what is called an *anticodon*. The sequence of bases in that anticodon might be, for example, adenine-uracil-guanine, or AUG. Now suppose you wanted to match this AUG anticodon

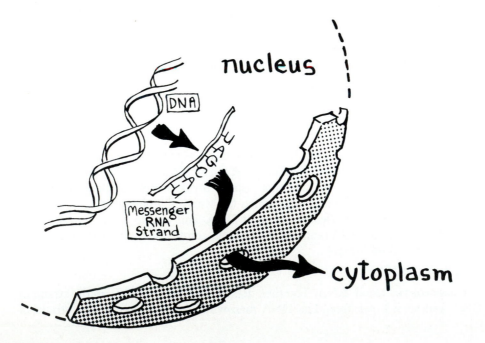

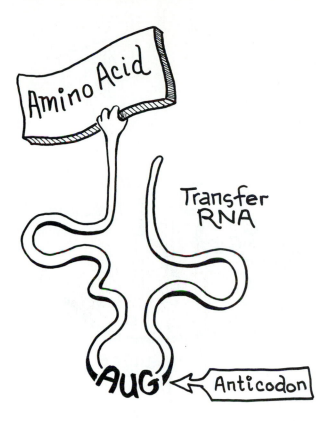

with a complementary sequence on a messenger RNA molecule. What sequence of nucleotides along the messenger RNA would provide this complementarity? I'll give you a hint. The first one would be—and I'll bet this will be a little surprise—uracil. Some of you thought it would be thymine. Remember, this is a matchup to a strand of RNA, not DNA, and RNA substitutes uracil for thymine. Okay, what are the next two?

STUDENT: Adenine and cytosine.

FARNSWORTH: Excellent. A transfer RNA with an AUG anticodon would bounce along the messenger RNA molecule until it stumbled onto a UAC sequence, and because all three nucleotides would be complements, they could bond up. A transfer RNA with a different anticodon, say GCC, would latch onto the messenger RNA at a different place. Now, I think you can see that the sequence of nucleotides along that messenger RNA is going to determine which transfer RNAs hook up to it and where they go. Since the transfer RNAs each carry with them an amino acid, I think you can see now how the sequence of nucleotides along the messenger RNA is going to say which amino acids are going to hook up to make our protein. Every three-nucleotide stretch of messenger RNA will have a complementary transfer RNA bonded to it, which in turn will bring along a specific amino acid. Okay, I see way in the back a question. Yes?

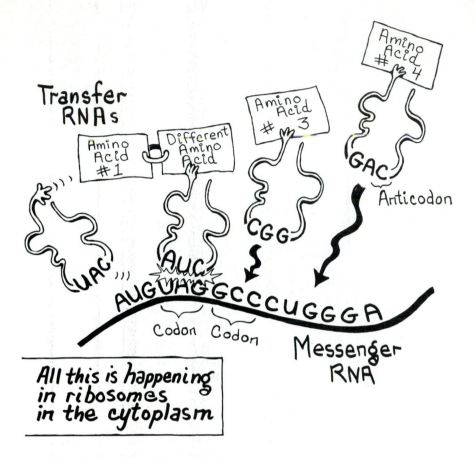

Transfer RNAs

Amino Acid #1

Different Amino Acid

Amino Acid #3

Amino Acid #4

GAC

Anticodon

UAC

AUC

CGG

AUGUAGGCCCUGGGA

Codon Codon

Messenger RNA

All this is happening in ribosomes in the cytoplasm

STUDENT: What happens between the amino acids that get brought in by the transfer RNAs? You've told us what happens between the transfer RNA and the messenger RNA; they get bonded together. What happens between the amino acids?

FARNSWORTH: Good question. They get bonded together, too, by using enzymes to form a peptide bond. That's what the final product of this whole process is, a polypeptide.

To recap a bit, we have this long messenger RNA, which is a complement of one of the DNA strands, out in the cytoplasm. The messenger RNA is long enough to contain the information necessary for one polypeptide chain that the cell needs. Transfer RNAs, each attached to a specific kind of amino acid, bounce along the length of the messenger RNA until the anticodon part of the transfer RNA finds a complementary three-nucleotide section of the messenger RNA; this section is called a *codon*. The transfer RNA bonds to the messenger RNA, and then as soon as one of these transfer RNA amino acid complexes gets a neighbor attaching to the next codon in line on the messenger RNA, the amino acids bond.

Now, this bonding process turns out to be not so simple. You don't just have those two amino acids saying "howdy" and grabbing on to each other. To get them to bond together, you need a special kind of cell structure called a *ribosome*. The ribosome acts kind of like a two-part mold. You put one part over the messenger RNA, the other part over the amino acids you want to bond together, then squish the two parts of the ribosome together. (*He cups his hands together.*) There are a whole bunch of enzymes inside the ribosome that catalyze the binding of the amino acids to each other. The ribosomes move down the messenger RNA molecule, bonding the amino acids one at a time.

So now you have these ribosomes bonding amino acids until you finally have the protein you want. After the protein is synthesized, the messenger RNA is usually broken up into nucleotides again, but in some cases, you can make many copies of a protein off a single messenger. The whole process of making the protein off the RNA is called *translation*. Okay, questions?

STUDENT: How does the DNA know where one RNA for one protein ends and another one begins?

FARNSWORTH: Hoo, boy, we got ourselves a molecular biologist here. I could answer you right now, but I think I can give you a better answer if you let me stall it off for a couple of minutes, okay?

Well, one of the biggest problems the early molecular biologists faced was this: We know that the sequence of nucleotides along a DNA molecule somehow tells the cells which amino acid goes where when you make a protein. We wouldn't have a problem if there were 20 different DNA nucleotides to match the 20 different amino acids, but there aren't; there are only 4. So at any one spot along a DNA molecule, you could have only 1 of 4 possible nucleotides, but at any given spot on a protein, you could have 20 possible amino acids. There is not a one-to-one relationship between the kinds of nucleotides and the kinds of amino acids.

I can give you an example of the kind of problem this is. Suppose you had a company in which all the instructions for making its product were kept in an office in a country called Nucland, whereas the production line was in a country called Ribosomia. Naturally, the guys in central control in Nucland don't speak the same language as the Ribosomians, so what you have to do is take the instructions, which are in Nuclandian, make a copy of them (to keep the originals safe), send the copy to Ribosomia, and then translate them from Nuclandian to Ribosomian.

So far, so good, but Nuclandian is a very weird language. It has only four letters; they are pronounced "adenine," "guanine," "cytosine," and "thymine" and written *A*, *G*, *C*, and *T*. But Ribosomian is different. How many letters would you guess are in the Ribosomian alphabet?

STUDENT: Twenty.

FARNSWORTH: Cookin' today. Twenty, and if you thought the Nuclandian letters were weird, try the Ribosomian ones. I won't go through all of them, but they're pronounced "alanine," "aspartic acid," "lysine," and so forth, and they're written *Ala, Asp, Lys*, etc.—you get the picture. So we can't just say one letter in Nuclandian equals a different letter in Ribosomian; there are only four different Nuclandian letters, but there are twenty different Ribosomian letters. But how about if we *combined* some Nuclandian letters to make Ribosomian equivalents? Maybe

Nuclandian

Letters: A, T, G, C.

Ribosomian

Letters: Phe, Leu, Ile, Met, Val, Ser, Pro, Thr, Ala, Tyr, His, Gln, Asn, Lys, Asp, Glu, Cys, Trp, Arg, Gly.

something like GC or AT. What could we call these combinations? Not words, because they represent only a single letter in a different language. Maybe we could call them "codons," because they're like a Morse code, or the binary code a computer uses—symbols that represent letters. Now, let's see, could we use a two-letter Nuclandian codon to equal a Ribosomian letter? But wait—before you answer, how many possible two-letter English words are there, not necessarily ones that make sense?

STUDENT: I don't know the number, but it would be 26 times 26.

FARNSWORTH: Six hundred seventy-six, just right. Each of the 26 possible first letters could match with 26 possible second letters, for a total of 676. But we're interested in Nuclandian. How many possible two-letter Nuclandian codons like *AT*, or *GC*, or *AA* could we have?

STUDENT: Sixteen.

FARNSWORTH: Right on. But we're still in trouble. We can say, for example, that *GC* translates into *Ala*, and we can assign 15 other Nuclandian two-letter codons to 15 other Ribosomian letters, but we're still shy 4 possible codons—we need 20. Okay, what happens if we go to three letters? How many three-letter condons like *ACA*, or *GCC* could we have?

STUDENT: Four times four times four.

FARNSWORTH: Well, your reasoning ability is better than your calculating ability, but that's right. Sixty-four different Nuclandian codons. We're golden now. We can say that GCA = Ala and translate all the other Ribosomian letters the same way, and we still have 44 possible codons left over. We know we will have to use a couple of those Nuclandian codons for punctuation. Why is this important? Look at this sequence of words. (*He writes "Don't Stop Don't Stop Don't Stop" on the board.*) There is an enormous difference between this—(*He writes "Don't. Stop. Don't. Stop. Don't. Stop."*)—and this—(*He writes "Don't stop. Don't stop. Don't stop."*). You see? Punctuation entirely changes the meaning. So we will use at least two of our codons to say, "Start the product with this Ribosomian letter" and "Stop the product with this Ribosomian letter."

We still have a bunch of Nuclandian codons left over, so what we will do is use these extra codons to add *redundancy* to our system in order to reduce the chance of error. Remember, we're making a copy of the original instructions in the office in Nuclandia to send to Ribosomia; we don't send the original. Suppose the code clerk makes a mistake. Let's say he wants to say "put Ala at this point." Ala in Nuclandian is CGC, but the clerk is all buzzed out on some controlled substance and writes "CGA" instead. If there were only one Nuclandian codon for each Ribosomian letter, we'd be shafted. However, *CGC, CGA,* and *CGG all* mean "Ala," so we're saved. If the clerk had written "AGC," instead of "CGC," that *would* have

caused an error. Redundancy can reduce the chance of error, but it can't eliminate it altogether.

Okay, let's leave the happy land of Nuclandia and go back to the cell. You don't have to be the world's foremost genius to see the direction this metaphor was taking. There are four letters in the DNA "alphabet," but there are 20 in the RNA alphabet. After decades of experimenting, it was found that a three-letter "codon" is sufficient to specify the type of amino acid at a particular spot on a polypeptide chain that is being synthesized. The arrangement of three-letter codons is called the *genetic code*, and each codon is called a *triplet*.

Now, where's the guy who asked the question about starting and stopping transcription? (*He spots the student.*) Okay, you've been given a partial answer. Some of the codons are start and stop signals. But there is more to it than that. It has been found that certain lengths of DNA contain the information that tells the cell *how* to make a particular protein. This length of DNA is called a *structural gene*. But there are other lengths of DNA, usually found right next to the structural gene for a particular protein, that tell the cell *when* to make the protein, and *how much* of it to make. These lengths of DNA are called *regulatory genes*.

There's another thing about this process you should know. Let's say you have a structural gene that contains the instructions for a protein that is, say, 100 amino acids long. How many codons would you expect to find in that gene, not counting the start and stop codons? A hundred, right? Wrong. There will be more codons than you need to make the protein. So you unwind your DNA and transcribe a messenger RNA that has these extra codons. The lengths of codons that aren't going to be used for structural purposes are called *introns*. The ones that are to be

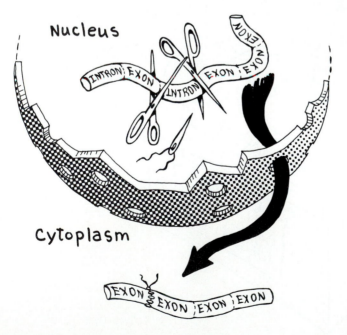

used are called *exons*. Anyway, before this messenger RNA leaves the nucleus, somehow the cell slices out the introns and splices the exons together. This is called *RNA splicing*, appropriately enough. The messenger RNA is then sent out into the cytoplasm and goes about its business. Now you're going to ask, "What do these introns do; aren't they excess baggage, molecular garbage?" The answer is that we don't know what they do. There is however, the intriguing observation that in higher animals the introns are longer, and there are more of them, than in lower animals. What's the significance of this? Nobody really knows; maybe you guys will be the ones who find out.

All right, let's tie this together. In the nucleus, you have the DNA. When you need to make a protein, you unwind part of the DNA and transcribe a length of messenger RNA. You snip out the introns, splice the rest together, and ship the messenger out into the cytoplasm. There, transfer RNAs carrying specific amino acids bond to the messenger RNA, and the ribosomes bond the amino acids. In effect, you have then translated the sequence of nucleotides on the DNA into a sequence of amino acids on a protein. This translation is done using the genetic code, which says that a certain sequence of three nucleotides on the DNA specifies a particular amino acid.

(*He steps away from the podium and the overhead projector.*) Now, this molecular business is extremely important, and you can't say you understand biology unless you understand it, but A'hm jes' an ol' country scientist, an' Ah likes to *see* what A'hm astudyin', so when we get together next time, we'll start to talk about things, like cells, that might be a little easier to visualize. I hope so, anyway. (*He picks up his notes and exits, trailing a string of students asking about a forthcoming examination.*)

Day 10

Mitosis and Meiosis

Professor Farnsworth enters wearing a black turtleneck sweater and black jeans.

FARNSWORTH: Good morning. Complete change of pace today. We're going to move on to a different level of organization, and I think it might be a little easier for you to visualize. So far, we've mostly been talking about things at the molecular level. Fascinating, but you have to use your imagination. Now, we will be talking about things at the cellular level. You can actually see what I'm talking about through the microscope.

Okay, on to the subject of cells. I'm going to refer you to your text for descriptions of cell structures and functions; you shouldn't have any problem with them. One thing that does usually require a little explanation is cell reproduction, and that is what the lecture for today is about.

You basically have two kinds of cells: those that live by themselves and those that are part of a multicellular organism. Both kinds of cells have to reproduce themselves, and both do it in essentially the same way. The general process of cell division is called *mitosis*. There is another kind of special cell division found only in sexually reproducing animals that we'll talk about later in the lecture.

Here you have a generalized kind of cell. (*He places a transparency of a cell on the overhead projector.*) It has a nucleus in the middle, and it has a kind of vegetable soup mixture of small cell parts called *organelles* outside the nucleus. The organelles, the structures that support them, and the fluid they're in are collectively called the *cytoplasm*. Suppose, now, I wanted to make two cells out of this one. Could I just tie a string around the middle of the cell and then pinch the two halves apart? That would be okay for the cytoplasm, but it wouldn't do for the nucleus. There's only one nucleus here, so we have two choices—we could assign it to one the new cells, leaving the other without a nucleus, or we could split it in half, just as we split the whole cell. But there are some real problems with just splitting it, and here's why.

Inside the nucleus are some structures called *chromosomes*. *Chromosome* means "colored bodies," referring to the fact that if you stain the cell with dye so it's easier to see under the microscope, the chromosomes suck up the dye and are correspondingly brightly colored. These chromosomes contain the DNA in the nucleus. Now, before we go any further, there's something you have to get straight. You've come

146

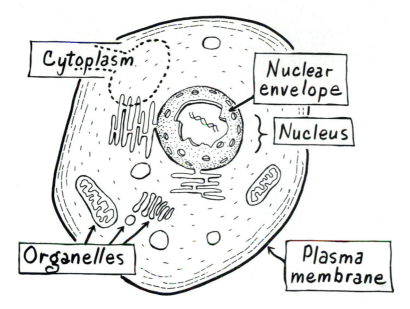

to understand the DNA molecule as being long and kind of stringy, right? Well, very shortly, I'm going to show you some pictures of chromosomes, and you're going to see that they can be long and stringy. I know what you're going to say— "Aha! This chromosome he's talking about is a DNA molecule." No! There is DNA *in* a chromosome, but there is also something else. In this case, a picture is worth a thousand words. (*He places a picture of a chromosome drawn at various levels of magnification on the overhead projector.*) In the picture you can see a double-stranded DNA molecule. We use this strand to wrap up a cluster of little balls of a protein called **histone**. The wrapped ball is a **nucleosome**. Using the same DNA strand we used to wrap the histone balls, we link together the nucleosomes and twist the strand of connected nucleosomes into a tight coil. In turn, we make a larger coil out of our tightly wrapped coil. What do we have now? Coils within coils within coils. Kind of a hypercoil.

When the cell is not dividing, this hypercoil strand of DNA and histone is loosely distributed all through the nucleus and is called the **chromatin network**. It would be sort of like having a piece of cooked macaroni that was maybe 100 meters long and throwing it in a swimming pool. It would just distribute itself more or less loosely and randomly in the pool.

When the cell starts getting ready to divide, this hypercoil starts to fold up on itself and wrap around itself, so that you eventually get something that looks like a Polish sausage under the microscope. This is the chromosome. Its structure is sort of like what would result if you took all that macaroni in the swimming pool, stuffed it in a long garbage bag, then removed the bag after the macaroni had dried out and gotten hard again. It would *look* solid but would really be this hollow tube wrapped around itself.

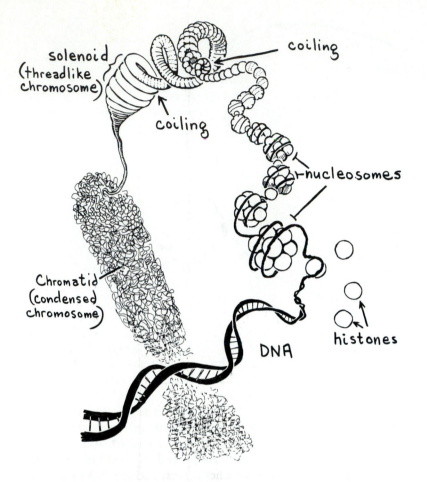

solenoid (threadlike chromosome)

coiling

coiling

nucleosomes

Chromatid (condensed chromosome)

DNA

histones

Now that we know what chromosome structure is like, we're ready to start dividing cells. Okay, here's a cell. (*He draws a cell on a transparency on the projector.*) Cell membrane, cytoplasm, nuclear membrane. This isn't a plant cell or an animal cell—it's one of these mythical average cells, like the so-called average person. Have I forgotten anything? Of course; chromosomes. How many chromosomes? Somebody pick an even number between 2 and 10—fine, 4. How many chromosomes do real organisms have? Two generalities: all the cells except the sex cells in a given kind of organism will have the same number of chromosomes, and different kinds of organisms can have different numbers of chromosomes. Humans have 46; some flies have four. Our imaginary cell has four also.

This is a cell that is not dividing, and not getting ready to divide in the immediate future. When it is in this state, it is said to be in *interphase*, which is kind of a dumb name because *interphase* means "between phases," but interphase *is* a phase. We old-timers use the word all the time, though, so you're stuck with it, I'm afraid.

MITOSIS

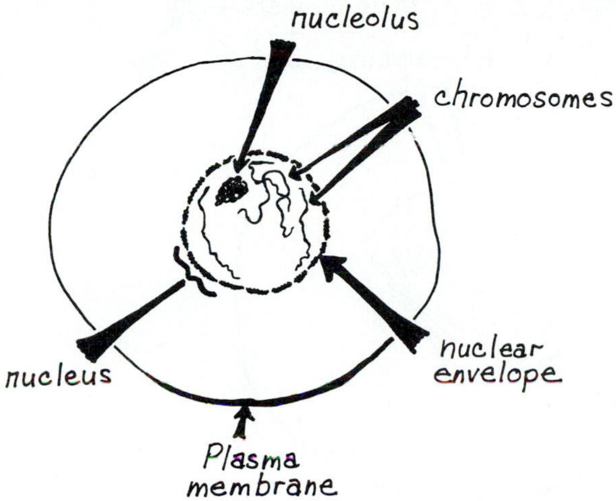

nucleolus

chromosomes

nuclear envelope

nucleus

Plasma membrane

Interphase

Anyway, in interphase, the cell goes about its business, doing whatever it is that it does: transmit nerve impulses if it's a nerve cell, contract if it's a muscle cell, and so on. In interphase, the chromosomes are stretched out so much that you can't really even see them under a light microscope. As the cell starts getting ready to divide, the chromosomes start to wrap up on themselves, and when they've condensed to the point where we can just begin to see them under the microscope, they look very much like thin macaroni. However, by the time we can see them, they have already replicated themselves. First, the DNA molecules in the chromosomes make copies of themselves, in the way that we talked about a couple of lectures ago. Then the chromosome itself replicates. So by the time you can see it, each chromosome has become a pair of *chromatids*. Our sample cell had four chromosomes, so now we have four *pairs* of chromatids. The sister chromatids—and don't ask me why they aren't called *brother chromatids*—are joined together at points on the chromatids called *centromeres*.

Time passes, and the chromatids shrink and condense further, until we get something that looks like a long, squiggly *H*, the crossbar of the H being the centromeres. At about this time, a couple of other things are happening. A cell structure called the *nucleolus*, which is very visible in the nucleus during interphase, starts to disappear, and the nuclear envelope starts to break apart. The cell is in *prophase* when these events occur.

MITOSIS

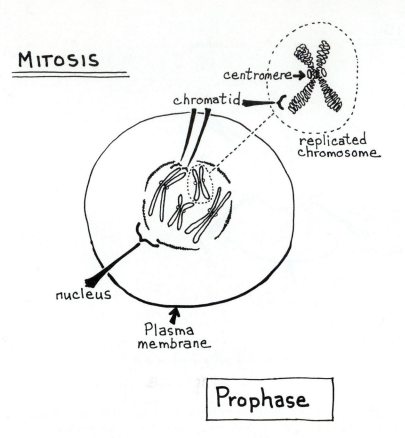

Prophase

I'll be talking about these various phases, but I want to make it clear that cell division doesn't proceed in a series of jumps; it's actually quite a smooth process, but these phases represent important points along the way. It's sort of as if we had only three pictures left in a camera and were trying to show what happened in a 100-yard race. We'd probably take one picture at the start, one in the middle, and one at the end. Same thing with phases; it's a continuous process, but the points we call *phases* are just important places along the way.

When the pairs of chromatids have shrunk up to nice, fat, little hot-doggy looking things, they start to move toward the middle of the cell, where they line up next to each other. When the chromatid pairs are aligned like a row of little H's lying on their sides, the cell is in **metaphase**. In the transition between prophase and metaphase, while the chromosomes are moving toward the equator of the cell, a rather striking thing happens. All of a sudden, we begin to see a bunch of very thin fibers, which stretch from points at opposite ends of the cell toward the chromosomes. The points at the ends of the cell where the fibers attach are called **spindle poles**. In animals, there is actually a physical structure at this pole called the **centriole**, where one end of each fiber connects.

MITOSIS

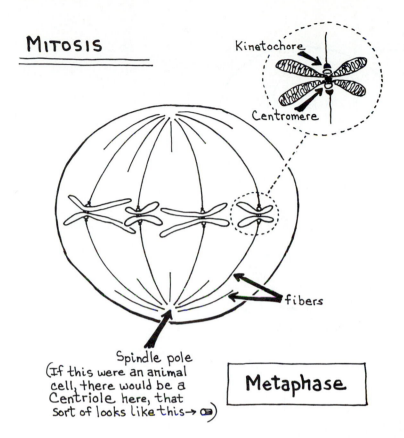

Kinetochore

Centromere

fibers

Spindle pole
(If this were an animal
cell, there would be a
Centriole here, that
sort of looks like this→ ▱)

| Metaphase |

At the chromosome end, a fiber attaches to each chromatid at the centromere point by a structure called the **kinetochore**, which acts like a kind of anchor for the spindle fibers.

Let me summarize what you have now. You've got your four pairs of chromatids, lined up next to each other at the equator, or middle, of the cell. At the centromere of each chromatid—and we've got eight chromatids in all, right?—you have a spindle fiber attached by a kinetochore. At the other end of the fiber, the fibers are gathered together at a common point, the spindle pole, one at each end of the cell.

At this point, something interesting happens. The whole cell starts to stretch out and pinch off in the middle. What does that do to the poles? It moves them away from each other. Since the fibers aren't stretchy, the pairs of chromatids get pulled away from each other, with the arms of the chromatids trailing behind the centromeres. Under the microscope, each separated chromatid would look like a *V*. Naturally, we wouldn't be good biologists if we didn't rename these separated sister chromatids. Now we're going to call them **daughter chromosomes**, another terrific, misleading name. It would be less confusing if we called them *offspring chromosomes*, but none of the books do that, so we're stuck. This step where the chromatids pull away from each other is **anaphase**.

Mitosis

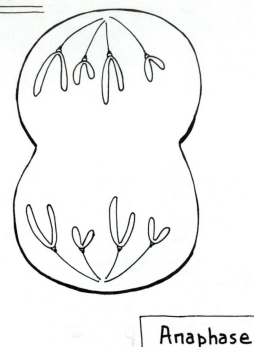

Anaphase

This would be a good time to review this chromosome terminology, because it always causes distress. We started with four chromosomes, right? Then each chromosome replicated during interphase, so we had four replicated chromosomes, each of which was made of two sister chromatids, giving us eight chromatids. When the sister chromatids separated just after metaphase, we started calling them daughter chromosomes, so now we have eight chromosomes. But wait—four of those daughter chromosomes are headed toward one pole, and the other four are going to the other pole.

It is during anaphase that the biggest remaining mystery of cell division occurs. We don't yet really know what makes the chromosomes move around in the cell. It is possible that the spindle fibers contract, like a rubber band, pulling the chromosomes toward the pole. It is also possible that the fibers are reeled in like a fishing line. There is evidence for both ideas, and maybe one of you will devise the crucial experiment that will resolve this.

Going back to the cell itself, it is elongating while contracting at the middle, forming a *cleavage furrow* around its waist. This step is called *telophase*. As the chromosomes get pulled closer and closer to their respective poles, a new nuclear envelope begins to form around each bunch of four chromosomes, new nucleoli appear, and finally, the two new cells completely pinch apart, yielding two brand new cells. Now, a question for you. Where is the original cell now?

Mitosis

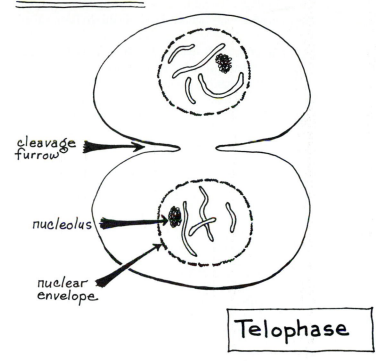

cleavage furrow

nucleolus

nuclear envelope

Telophase

The class seems puzzled.

FARNSWORTH: Curious, isn't it? It has disappeared, and two new, identical cells have replaced it. We had four different chromosomes in our original cell. In each of the daughter cells, we have four different chromosomes, which are the exact duplicates of the ones in the original cell.

There is a kind of parallelism between the molecular level and the cellular level here. In DNA, we start with a single double-stranded molecule, replicate it, and have two identical double-stranded molecules which, if everything goes right, are exact duplicates of the original molecule, which no longer exists. In the cell, we have a complement of chromosomes which replicate themselves; then each of the replicated chromosomes goes to a new cell, giving us two identical sets of chromosomes which are exact duplicates of the original chromosomes, which no longer exist.

I'm now going to tell you the single most important fact about mitosis, and that is that the number and kinds of chromosomes are preserved, generation after generation. You could have 100 divisions, yielding millions of cells, and each of the descendants of the original will have the same number and kinds of chromosomes as the ancestral cell. Furthermore, barring accidents, the information contained in the descendant chromosomes is the same as that in the original.

Mitosis is a beautiful, relatively simple, and elegant process, but it is not adequate for every cell division job in biology. There is a special situation involving sexually reproducing animals that calls for a specialized kind of cell division called *meiosis*.

Meiosis is used to produce the *gametes*, or sex cells, which in females are the *eggs* and in males the *sperm*. Why is a special process needed here? Think about it for a second. Let's say I'm an animal which has 10 chromosomes in each of its *somatic*, or body, cells. Well, sexual reproduction involves *fertilization*, or the fusion of gametes of opposite sex. Suppose I'm a male, I have 10 somatic chromosomes, and I start to produce sperm by mitosis. How many chromosomes will one of my sperm have?

STUDENT: Ten.

# of Chromosomes

(10) Somatic cell

(10) Somatic cell

MITOSIS

10

Sperm

MITOSIS

10

egg

20

Zygote

A System which doesn't work!

Production of gametes by MITOSIS results in twice as many chromosomes as in the parents — this would double every generation!

FARNSWORTH: Right. And my sweetheart, if she produced eggs by mitosis, how many chromo—?

STUDENT: Ten also.

FARNSWORTH: Right again. And when a sperm and egg fuse, how many chromosomes will the *zygote,* or fertilized egg, have? Rhetorical question—20, of course. And when that offspring grew up and became sexually mature, it would have 20 chromosomes in its gametes. There would be a doubling of chromosome number every generation, until after only a relatively small number of generations, you'd have millions of chromosomes in each gamete. Impossible situation. We have to devise a method of cell division that will cause the gametes to have *half* the number of chromosomes as the somatic cells. Then, when they fuse at fertilization, the zygote will have the somatic cell number of chromosomes and can then divide by mitosis until the organism grows up and starts producing gametes of its own. Okay, vocabulary time now. We call the number of chromosomes in a somatic cell of a particular organism the *diploid,* or *2n,* number. The number of chromosomes in the gametes is the *haploid,* or *1n,* number. So what is the haploid number for humans?

STUDENT: Twenty-three.

FARNSWORTH: Right. Half of 46. Each sperm cell will have 23 chromosomes, as will each egg. Fuse an egg and a sperm, and you get 46 again. Okay, now I'm ready to make a confession to you. I lied. I said humans have 46 *different* chromosomes. They don't. They have 23 *pairs* of chromosomes. Look at it this way. I stand before you, and every one of my cells has 46 chromosomes. Go back in time to when I was 5 years old, 1 year old, a fetus, an embryo, a *zygote.* How many chromosomes did the zygote have? Forty-six. And before the zygote was a zygote, what was it? Two gametes, each with 23 chromosomes. Not just *any* chromosomes, though. This is critical. Each gamete contributes one member of each pair of chromosomes the zygote has.

So, now we know that each gamete carries one member of each pair of chromosomes that the zygote and, later, the adult organism have. So if the diploid number, that is, the number of chromosomes that the zygote has, is 12, there are 6 pairs. That means that the gametes are each going to have to contribute, not just any six chromosomes, but one member of each of the six pairs.

What we need, then, is a kind of cell division in which a cell divides into two new cells, but instead of yielding daughter cells having the same number of chromosomes as the parent, as was the case in mitosis, it yields daughters with *half* the chromosome number of the parent. Just as importantly, that half will not just be any randomly selected half, but will be composed of one of each of the chromosomes making up the pairs. We'll call the members of a pair *homologous chromosomes*.

This chromosome-number-reducing process of cell division is called meiosis and occurs only in the *gonads,* or sex organs—*testes* in the male and *ovaries* in the female. Not all the cells in these organs undergo meiosis at the same time; if that happened, all the cells in the organs would get converted to gametes in one shot, and we don't want to do that.

The basic process operates in a similar manner for both eggs and sperm, and I'll point out the differences as they come up. So let's work an example now with a cell whose 2n number is, let's say, six.

I'm going to show you only what is happening in the nucleus here; the rest is about the same as in mitosis. We have three kinds of chromosome pairs: short ones, long ones, and squiggly ones. Now, where did these chromosomes originally come from? One member of each pair, the paternal one, came from a sperm; the other, the maternal one, came from an egg. In our figure, the ones that came from an egg are darkened in. We start out just as in mitosis; each chromosome replicates. Now we have three pairs of replicated chromosomes. They shrink up, the nuclear envelope disappears, and we're now in the meiotic prophase. The chromosomes start to move toward the middle of the cell, just as in mitosis, but there then follows an absolutely critical difference.

In mitosis, all the replicated chromosomes line up next to each other, side by side. In meiosis, the pairs of replicated homologous chromosomes line up *opposite* each other, across the middle of the cell. (*He places a new figure on the projector.*) I want you to notice something here in the figure. The maternal short chromosome is lying on one side of the equatorial plate, aligned with the paternal long one and then the paternal squiggly one. Is that the only way I could have shown this? Could it have been that the paternal short one could have been next to the paternal long one, and it in turn next to the maternal squiggly one? Yes. Are there other possible ways these homologous chromosomes could have lined up? Yes. Now, what do you think determines which way they *actually* line up in a real cell? The answer is chance. Pure, random chance. Each pair of homologous replicated chromosomes could line up with either the maternal or

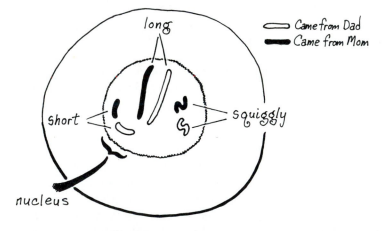

This is a cell which is going to
        undergo **MEIOSIS.**
At this point it looks and acts like a cell
which will divide by MITOSIS.

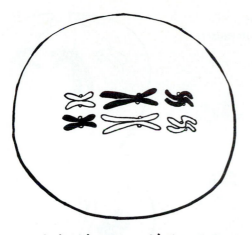

First Meiotic Metaphase

This is different from Mitosis,
where the replicated chromosomes
will have lined up something
like this

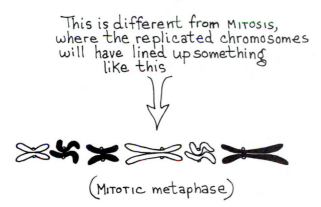

(Mitotic metaphase)

the paternal chromosome on a given side of the cell's equator. This is going to be an extraordinarily important observation in just a moment.

We are now at the first meiotic metaphase. We get fibrils and kinetochores just as in mitosis, and the cell starts to stretch out just as in mitosis, but here's the big difference. Instead of having all the chromosomes lined up side by side, and the sister chromatids getting pulled apart, the *replicated homologous chromosomes get separated*. This is the first meiotic anaphase. Then the cell finishes what we call the *first meiotic division*. Now we have two cells, and each cell has one replicated homologous chromosome of each kind of chromosome pair, in this case, short, long, and squiggly. Are the two cells *exactly* alike? No. One has a maternal short, a paternal long, and a paternal squiggly chromosome. The other has a paternal short, a maternal long, and maternal squiggly chromosome. If the chromosomes had lined up differently at metaphase, by chance, we might have gotten different combinations of maternal and paternal chromosomes. So what happens next?

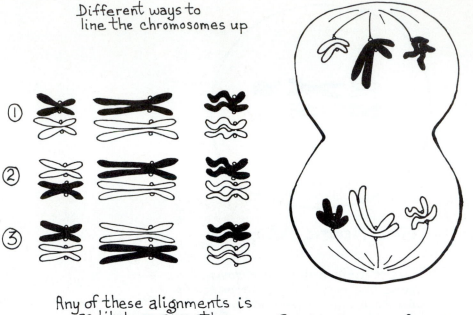

Different ways to line the chromosomes up

① ② ③

Any of these alignments is as likely as any other

First Meiotic Anaphase

Each daughter cell then divides in normal mitotic fashion. The sister chromatids get pulled apart just as in mitosis. What we have now is *four* cells, when originally we had one. How many chromosomes does each cell have?

STUDENT: Three.

FARNSWORTH: And how many chromosomes did the original diploid cell have?

STUDENT: Six.

FARNSWORTH: Excellent. We have now reduced the number of chromosomes in our soon-to-be gametes to the haploid number. If this were a male, each of these four cells would develop into an individual sperm. If this **gametogenesis** (*genesis* being "beginning of" or "origination of") were occurring in a female, there would be a difference. After the first meiotic division, one of the two cells would normally die and become a **polar body**, leaving the other one to undergo the second meiotic division. After the second division, one of *those* two cells would die, also becoming a polar body. So one male cell that starts meiosis becomes four sperm, but one female cell that starts meiosis normally becomes just one egg.

Okay, that's the process. Now let's look at some of its implications. Meiosis provides a means of halving the number of chromosomes in the gametes, but equally important is the fact that meiosis provides an opportunity to *scramble* the chromosomes, providing variety among offspring that develop from zygotes. This makes sense. The chromosomes contain DNA, which contains information about what your physical characteristics, like eye color, will be. Let's say the information for

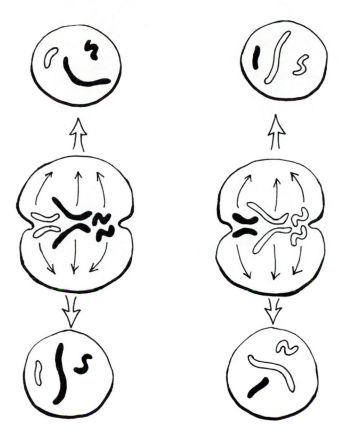

Notice that you have 4 possible gametes
    of 2 different kinds

eye color is on a particular pair of chromosomes. You got one of those chromosomes
from your mother, the other from your father. Consider now the one from your
father. Where did he get it? From either *his* mother or his father. And what determined whether he passed his mother's or his father's eye color chromosome on
to you? Chance. Chance at the time the chromosomes lined up at the equator of the
cell during meiosis, and chance during fertilization. If another of your father's sperm
had connected with your mother's egg, you might have had quite different characteristics. This chance factor, and the mechanism of meiosis, is why brothers and
sisters don't look exactly alike, even from the same mother and father, and why you
might have your father's mother's eyes, your mother's mother's chin, your father's
father's eyebrows, and so on.

   Let me sum this all up with one giant diagram here. (*He places a new transparency
on the overhead projector*). We could start anywhere, but we'll start with a sperm and
an egg, each having three chromosomes. They fuse at fertilization to form a zygote
with six chromosomes; three pairs. The zygote divides by mitosis until you have an

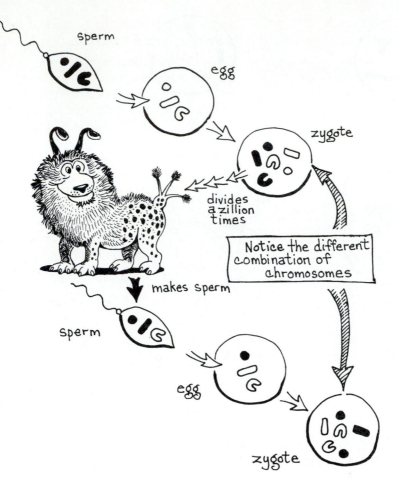

adult organism, each cell of which has six chromosomes. In the gonads of this organism, let's say it's a male, some of its cells undergo meiosis and make sperm. Recall that the process of meiosis produces four sperm, two each of two different chromosome combinations. The organism is actually producing millions of sperm from meiosis that is going on in millions of cells. We'll follow only one of these sperm. Notice that the chromosomes in this sperm are different from the ones we started with up at the top of the drawing.

Okay, this male gets together with a female, fertilization takes place, and we get a new organism. The old organism can now go ahead and die if it wants to; it has transmitted its chromosomes. The chromosomes themselves don't change from generation to generation, but the combination of chromosomes does. Any questions now?

STUDENT: Yeah. In humans, I can see where you could get millions of different combinations of 46 chromosomes, but when you look around the room, people pretty much look alike. We don't have humans that look like part radish, part chicken, part human. I don't see that there really is that much variation.

FARNSWORTH: Good observation. The range for *each* characteristic is not too big. It is the combination of *different* characteristics that is big. Look at eye color. You have a relatively narrow range. Blue, brown, maybe a couple of odd ones—greens and purples. No red ones. No yellow ones. Let's just say there are four different kinds of eye color. Hair color. Black, red, blonde, brown. No purple. No pink. No green. Really kind of boring. But if we say there are four kinds of eye color, and four of hair color, how many different combinations of hair and eyes are there? Sixteen. Now, let's say there were only 40 identifiable human physical characteristics. Obviously, there are many more than that, but let's just pick that number. And let's further say that each characteristic can vary in only four ways, how many total different combinations could you have? (*He pauses and thinks.*) Ah, it would be a 1 followed by 24 zeros, a huge number. But the variation in any *one* characteristic is limited to those possible variables that humans can have.

Good, so next time, we'll talk about some ways of predicting the likelihood that an offspring will have a particular set of characteristics. Remember, we have an exam coming up shortly and the time to start reviewing is now. (*He exits.*)

Editor's note: There is more disagreement between biologists about terminology in this area, than almost any other. No two basic books will have exactly the same definitions of mitosis, cytoplasm, centromere, chromosome, chromatin, and several other important terms. There is no real disagreement about *how* the process of cell division works, but there are historical reasons why there are different views about terminology. If you are a student using *Professor Farnsworth's Explanations* in association with a general biology textbook, and you find that the text uses a different terminology than you see here, don't be alarmed; just ask your instructor which terminology he or she prefers.

Day 11

Basic Mendelian Genetics

Professor Farnsworth enters wearing a leather vest and western shirt with pearl buttons. It is clear that his style is no style—he apparently wears whatever happens to be available at the moment.

FARNSWORTH: Good morning. Oh, are we going to have *fun* today. (*He rubs his hands together in mock glee.*) Yes, today, instead of just describing things, we're going to *do* them. We're going to get the brain unlimbered a little, do a few problems. First, though (*he walks out into the class and starts to look around*), I need—a redhead. Dum-te-dum, ah, here's one, and a good flaming red one to be sure. All right, now, tell me, if you please, who in your family has hair like that? Mother, father, grandmother?

REDHEAD: My grandfather on my mother's side. Mom has sort of red hair, but not like this.

FARNSWORTH: Excellent; my congratulations. Now, let's see, hmmm (*looks around the class*). Blue eyes. Yes, I need a pair of startlingly blue eyes, and (*stops by a student*) sure enough, here they are. Let me guess; your father has blue eyes, right?

BLUE EYES: Wrong. My mother.

FARNSWORTH: Ah well, close. (*He runs back up to the stage.*) Today, we're going to start talking about inheritance—how characteristics are passed from generation to generation. The science of inheritance is called *genetics*, and we're going to have two or three lectures exploring how characteristics are inherited.

You can't separate genetic studies from the facts of sexual reproduction, the most important of which is the observation that you have one gamete, a sperm, coming from a male and another gamete, an egg, coming from the female. Put 'em together and what have you got? A zygote, which grows up to be an adult and eventually produces gametes of its own.

For our purposes now, we can look at an egg and a sperm as information bearers. Information from the father and mother is combined in the zygote to produce the characteristics of the organism that grows up from the zygote.

You can think of an organism as a collection of characteristics or traits. Some of these traits will be easily observable, like hair or eye color. Others will be cellular

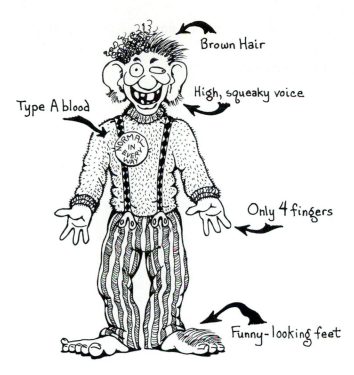

Brown Hair

High, squeaky voice

Type A blood

NORMAL IN EVERY WAY

Only 4 fingers

Funny-looking feet

traits, or biochemical traits, like blood type or the possession of certain kinds of digestive enzymes. Behavioral traits can even be inherited, for example, bird song. Some birds can sing a song characteristic of their species without ever having heard an example. So the whole organism is made up of thousands of different traits.

Some of these traits can demonstrate a wide spectrum, for example, skin color in humans. Other traits, like the presence or absence of hemoglobin, the compound that hauls oxygen around in the blood, have essentially no variety in a normal animal—either you have it or you die in the embryonic stage.

So what determines which characteristic *you* will have, of the range of possibilities for each trait? This question can be answered on two levels: the first, biochemical; the second, genetic. Look at hair color. What makes hair black? The presence of a pigment in the hair. What causes the pigment to be made? The presence of an enzyme specific for the production of that particular pigment. If you have the enzyme, you have the pigment; if you have the pigment, you have the hair color. No enzyme, no black hair. Same story for the less obvious traits. If you have one kind of enzyme, you have one kind of blood type; if you have a different enzyme, you have a different blood type.

Ultimately, then, all the thousands of characteristics you have are determined by the presence or absence of certain enzymes. And, someone remind me, what is an enzyme?

STUDENT: A protein.

FARNSWORTH: Ah, yes, and how does an organism know how to make this protein?

STUDENT: From information contained on a DNA molecule.

FARNSWORTH: We're on a roll. And where is this DNA found?

STUDENT: In chromosomes.

FARNSWORTH: *Wunderbar!* Now we're ready to start putting two and two together. The thing that determines which characteristic you have is the presence or absence of an enzyme. The information to produce the enzyme is in the chromosomes. Here's where we start to get genetic now. Remember that the organism has *pairs* of chromosomes? And one of the members of that pair came from the sperm, one from the egg. Okay, think back to meiosis now, and how chromosomes could be scrambled up so that some gametes have one member of the pair, and others have the other member.

All right, so now the egg has a homologous chromosome, and the sperm has a homologous chromosome. The sperm and the egg get together and now you have your homologous chromosomes together. Okay, here's the important part. The information needed to produce a particular enzyme, which in turn will produce a particular trait, is contained in a specific place on each of the two homologous chromosomes. The information from both the homologous chromosomes determines what characteristic you'll have.

Before we go any further, we need some vocabulary. If we're talking about the appearance of a particular trait, we call that appearance the **phenotype**. Blue eyes are a phenotype. The genetic information to produce that trait is called the **genotype**. That information exists on both homologous chromosomes. The information on any one of the chromosomes is called a **gene**; the information on both is a *pair of genes*. Genes can come in different forms, and each of the forms is called an **allele**.

This is a lot of vocabulary in a short space of time, so let's use it a bit. Here's a picture of a pair of homologous chromosomes, inside each of which is a length of DNA. (*He places a transparency on the overhead projector.*) This DNA contains the instructions for making many different polypeptides, most of which are enzymes. Remember that we called a length of DNA which contains the instructions for an enzyme a *gene* when we were talking about molecular biology? Well, now we see that that enzyme can determine whether you have a certain characteristic or not, so we can define a gene in a slightly different way now: a gene is a unit of information that can be used to specify a trait. It normally takes two genes, at the same place (called the **gene locus**) on a pair of homologous chromosomes, to define a trait for an individual. There are some special cases we'll see in a little bit in which only one gene, on one chromosome, can produce a certain phenotype.

Now, I said that genes come in different forms, or alleles. In the simplest situation, there are only two alleles possible, and they're usually represented by a capital letter, for example, like big B, and a small letter, like little b. In this simple situation, we call one of the alleles, the one we label with the capital letter, the

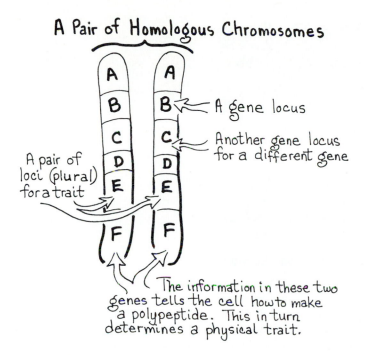

A Pair of Homologous Chromosomes

A gene locus

Another gene locus for a different gene

A pair of loci (plural) for a trait

The information in these two genes tells the cell how to make a polypeptide. This in turn determines a physical trait.

dominant allele and the lowercase one the *recessive* allele. I have to give you warning. These names, *dominant* and *recessive*, are left over from the old days of biology. We still use them, but they can be misleading. *Dominant* and *recessive* in genetics don't refer to some weird sex practice. We're not talking about "ruling over" or "being submissive" here.

Every individual has two genes for a particular trait. Exceptions exist for certain sexual characteristics. If there are two allelic forms of that gene, that means that there are several possible combinations of those two genes. If both of the genes are the dominant allelic form, we say that the individual has a *dominant homozygous* genotype—*homo* means "same." If we were going to symbolize this for a *B* gene, we would show a dominant homozygous individual as being *BB*. What this means is that on the pair of chromosomes that carry the two genes for the particular trait, the *B* allele is found on both. On the other hand, if one of those genes is a *B* and the other is a *b*, the individual has a *heterozygous*—*hetero* meaning "different"—genotype and is symbolized as a *Bb*. The final combination is called a *homozygous recessive* genotype; this kind of individual has two *b* alleles and is symbolized as a *bb*.

Okay, now, what is the relationship between what your genotype is and what your phenotype is? Well, sometimes there is a very simple relationship, and at other times it is so complicated it will make your head spin. Let's start with a simple case, and to make it even easier, we'll use an imaginary characteristic—werewolfism. If you and your spouse are werewolves, what's the likelihood that some fraction of your kids will be werewolves too?

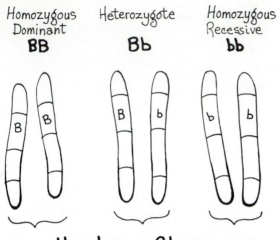

Homozygous Dominant **BB** Heterozygote **Bb** Homozygous Recessive **bb**

Homologous Chromosomes

As it happens, the gene for werewolfism operates according to the same set of rules that the founder of genetics, Mendel, discovered for pea seed colors; however, pea seeds are boring when you can talk about werewolves. The set of rules is called *complete dominance*, and it works like this—but first, we have to give a name to the werewolf gene; let's call it *W*.

The dominant werewolf allele is *W*, the recessive is *w*, so you could have three possible genotypes, right? What are they?

STUDENT: *WW*, *Ww*, and *ww*.

FARNSWORTH: Good. The dominant homozygote, the heterozygote, and the recessive homozygote. In complete dominance, if you have at least one dominant allele, you have what's called the *dominant phenotype* and in this case, you are a werewolf. If you don't have a dominant allele, you have the *recessive phenotype;* here, that means you are not a werewolf. So, now, what are the possible genotypes for a werewolf? Somebody in the middle, now. Yes, Geoff?

GEOFF: A werewolf could be a *WW* or a *Ww*.

FARNSWORTH: Fine. So that means a non-werewolf has to be a what?

GEOFF: A non-werewolf is a *ww*.

FARNSWORTH: Perfect. Now, just by looking at a werewolf, could you tell whether it was a dominant homozygote or a heterozygote? No. It could be either. Could you tell the genotype of a non-werewolf? Yes, because non-werewolves can have only the homozygous recessive genotype.

So far, so good, but here it is full moon, and down in the graveyard, two werewolves meet. They have a relationship and decide to get married, but they're concerned that some of their kids might be non-werewolves, which would be terrible

Werewolf   Non-werewolf

Could be WW or Ww   Has to be ww

for them, of course. As a genetics counselor, you want to advise them. What would be the first thing you would want to know, in order to help them?

STUDENT: You'd need to know their genotypes.

FARNSWORTH: That's right, but there's a hitch here. Remember, in the case of the werewolf gene, you can't just look at someone who has the dominant phenotype and know what their genotype is. But are we dead yet? Is there anything we could find out about them that might help us nail down their genotype?

STUDENT: Yes, the phenotype of their parents.

Puzzled looks from class.

FARNSWORTH: Yes, very clever of you to figure it out, but we need to know why that information is useful. It has to do with which allele goes into which gamete. Look at this. (*He puts a transparency on the overhead projector.*)

Suppose you are a heterozygous werewolf. That means that in all your cells, on the pair of chromosomes that has the werewolf genes, you have one W allele, on one chromosome, and a w allele on the other homologous chromosome. So now, comes time for you to make gametes. At meiosis, that pair of homologous chromosomes will go into separate gametes; one group of gametes will have a chromosome with a W allele, and the other group will have a chromosome carrying a w allele. So you have the potential of producing either a W or a w gamete.

But suppose you were a non-werewolf. You would be a ww, and that means that both the chromosomes carrying the werewolf gene have w alleles. You can produce only one kind of gamete, w.

Now, let's go back to our werewolves in counseling. Pick one of them, it doesn't matter which, let's say the female. All we know is that she's a werewolf—she could be either Ww or WW, and she'd still be a werewolf. But now, suppose either her

WW }
Ww } ?   ww

WW } ?   WW } ?
Ww }   Ww }

But— if either of her parents
is non-werewolf, she would
then have to be **Ww.**

father or her mother were a non-werewolf. Could we say anything about either her genotype, or the genotype of her other parent? Certainly.

We now know that one parent was a non-werewolf. What genotype would that parent have to be? A *ww*—otherwise he or she would be a werewolf. So that parent could contribute only a *w*-bearing gamete. How about the other parent now? Well, now we know that our werewolf counselee *has* to be a heterozygote, don't we—the normal parent could produce only *w* gametes. The *W* of our counselee could come *only* from her other parent, so that parent was definitely a werewolf, and could be either a *Ww* or a *WW*. To know which, we'd have to go back one more generation.

To recap, *if* one of our counselee's parents were a non-werewolf, we could say with absolute certainty that the counselee was a heterozygote. Okay, we do our little family history, and we find out that both our werewolf clients are heterozygotes. Now what do we need to know to tell them what the likelihood is that they might have a non-werewolf kid?

STUDENT: You'd need to know what kind of gametes they could produce.

FARNSWORTH: Exactly correct. Our job is made easier because both of them have the same genotype, *Ww*. Let's take the male first. He has the potential of producing *W*-bearing or *w*-bearing sperm. If we took a sperm sample from him—although I guess you don't *take* a sperm sample from a werewolf; you ask him for it politely— anyway, if we got a sperm sample and, at random, pulled out one sperm cell, which would it be more likely to be, a *W* or a *w?*

STUDENT: Neither—there'd be an equal chance of its being either one.

FARNSWORTH: Right. *W*'s and *w*'s would be produced in equal numbers, so any one sperm cell could be either a *W* or a *w*. The same would be true for her. She would be equally likely to produce a *W* or a *w* egg. Okay, now we know what kinds of gametes each of them can produce; they both can produce *W*'s and *w*'s in equal numbers. Now what do we need to do?

STUDENT: Uh, maybe you need to match up different kinds of sperm with different kinds of eggs to see what the genotypes would be of the offspring.

FARNSWORTH: Very good. The guy is producing millions of sperm, and what is the probability that any one of them will contain either a *W* or a *w?* Raise hands, how many think it's 50-50? (*Most students raise their hands.*)

Ha! Caught you! You have to listen carefully to genetics questions. I said what is the probability that it will be *either W* or *w*. It has to be one or the other, so there is a 100 percent probability that it will be either one. If I had asked what is the probability that any particular one would be a *W*, the answer would be 50-50. Anyway, he can produce both kinds, and it is equally likely that the sperm that fertilizes her egg will be either a *W* or a *w*.

The situation for her is similar. She has an equal chance of producing a *W* or a *w* egg. What we have to do now is match up all the different kinds of sperm he can produce with all the kinds of eggs she can produce. It's important to understand what question we're asking. We are asking, "What kinds of genotypes could potentially be produced among the offspring of parents with these given genotypes?" We are *not* asking what genotype a given offspring, say, their first child, *will* be; only what it *could* be. We're talking about probabilities and possibilities, not certainties and sure things.

There are actually a couple of ways to do this matching with pencil and paper. I'll show you first a way of matching them up with arrows. This is a good way to start because it forces you to realize that what you're doing is making trial combinations of different eggs and sperm. You make a column and call it the "male" column. Under it, you list the kinds of sperm the male can produce. Since we're talking about the werewolf gene here, we've already determined that he can produce *W*- and *w*-bearing sperm. Then you make another column, the "female" column, and you do the same thing—list the kinds of gametes she can produce. In this case, also *W* and *w*. Then, starting with the male column, you say "What kind of offspring would you get if a *W* sperm fertilized a *W* egg?" and you draw an arrow from the *W* sperm to the *W* egg. The *W* and the *W*, when they combine, would

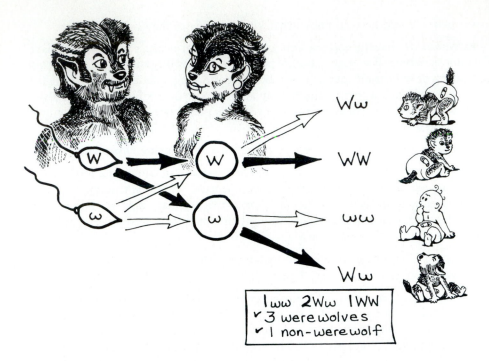

Iww 2Ww IWW
✓ 3 werewolves
✓ I non-werewolf

produce a *WW* offspring, which would be a werewolf. Then you do the same thing with all the other combinations, and you see that we have four possible combinations. Somebody, now, what are the genotypes of the offspring which would result from each of these combinations?

STUDENT: One *WW*, two *Ww's*, and a *ww*.

FARNSWORTH: Right. And how about the phenotypes?

ANOTHER STUDENT: Three werewolves and a non-werewolf.

FARNSWORTH: Good. Okay, let's talk about the implications of some of this. If they had a very large number of offspring, large enough so the rules of statistics would start to operate, say 10 or more, we would expect that about three-quarters of the offspring would be werewolves. If we were talking about genotypes, about half of them would be heterozygotes. But how about if we were talking, not about groups of offspring, but about a single offspring. Could we say for sure what any one offspring's genotype would be? No. We could talk only about the *probability* that it would have a particular genotype. In this case, since there is only one combination of eggs and sperm that would produce a *WW* offspring, but there are two combinations that will produce a *Ww*, the chances are twice as good that any one offspring will be a *Ww* than that it would be a *WW*. Since there are four possible combinations, and two of those combinations produce a *Ww*, if we asked the question "What is the probability that any one offspring will be a *Ww*?" the probability would be 2 chances out of 4. Another way of saying that would be 1 out of 2, or half and half—there's as much chance that it will be a *Ww* as that it won't be.

Okay, the arrow matching technique is good for simple problems like this, but in the future, when you're doing more complicated things, you need a better way of keeping track of combinations. A man named Punnett developed such a technique, and we call it a **Punnett box** today. You'll see this technique called a "Punnett square" in many books, but as we'll see, you produce a square only under certain conditions. All you do is start out just as we did before. Make a vertical list of the kinds of sperm, but instead of matching the eggs vertically, you list the kinds of eggs horizontally up above. Then you draw in boxes where the rows and columns intersect. You match the first sperm with the first egg, and put the genotype of the offspring in the corresponding box; then you do the same with all the other combinations. Easy, isn't it?

Okay, let me see if you know how to work this now. Suppose the male werewolf's a heterozygote, as before, but the females's a homozygous recessive. What kind of gametes can she produce?

STUDENT: Only *w*'s.

FARNSWORTH: That's right. So he has a *W* and a *w* on the vertical axis. She produces only *w* eggs, so you put the single *w* at the top of the column. Match up the

Monohybrid Cross

gametes, and what do you have? Only *Ww* and *ww* as possible offspring. So could this couple have a homozygous dominant offspring? No. If they had a lot of kids, what fraction of them could be expected to be werewolves? Half. Now this kind of cross, in which you're looking at one trait, which is determined by a pair of genes on one pair of chromosomes, is called a ***monohybrid cross***. Let me now go over again the general procedure for working out a monohybrid cross.

First, list the genotypes of the parents, if you know them. You might have to do a little family history first. Sometimes, you can't say for sure—as we saw, just by looking at a werewolf, you can't tell if it is a *WW* or a *Ww*. If that's the case, you assume that there's a fifty-fifty chance of its being a *WW*, because without having a family history or any other information, you start by saying there is just as much chance that an organism showing a dominant phenotype is a dominant homozygote as there is that it's a heterozygote. After you have pegged the genotype of the parents as best you can, you ask what kinds of gametes each parent can produce. Then, using a Punnett box, you match up all the possible gametes, which will give you the phenotypes and genotypes of the possible offspring. And that's basically all there is to it.

Now, before we finish up, there's one more thing I want to tell you about. So far, we have been dealing with a set of genetic rules called *complete dominance*, in which the phenotypes of the dominant homozygote and the heterozygote are the same. Other traits can follow a different set of rules, in which the dominant homozygote, the heterozygote, and the recessive homozygote all have different phenotypes. In a typical case, like flower colors, the heterozygote is intermediate in color between the two homozygotes—maybe the dominant homozygote is red, the heterozygote is pink, and the recessive homozygote is white. This set of rules is called ***incomplete dominance***. The only way you can find out whether a given gene operates by complete or incomplete dominance rules is by doing a series of crosses and looking at the offspring. Let's look at another imaginary example.

Hmmm, let's see. Yeah, you know how in all the old creature feature horror movies on television it was always something like *Wolfman Meets Dracula* or something like that? Well, that's because a lot of those characteristics are inherited by an incompletely dominant gene. The gene is called the *V gene*. If you are a homozygous dominant, a *VV*, you are a vampire; if you are a heterozygote, you are a zombie; and if you are a *vv*, you are, ummmm, oh yeah, now I remember, a ghoul. Problem time. If a male zombie marries a female ghoul, first, what fraction of their kids could be expected to be vampires, and, second, what fraction of their kids could be expected to have the *vv* genotype? (*He puts a transparency on the overhead projector and completes the Punnett box as the students work at their seats.*)

Answer time. Fraction of vampires?

STUDENT: A quarter of them.

FARNSWORTH: A good thought. Regretfully, a wrong thought. A vampire is a *VV*, and so will have to get a *V* from each parent. Ghouls, however, are *vv* and can't contribute a *V* gene. So a male zombie and a female ghoul have no chance of having vampire kids. Second part of the question: what fraction of their kids could be

expected to have *vv* genotypes? We're running out of time here, so I'll tell you—it would be half. As you can see now, an incomplete dominance problem is set up exactly the same way as a complete dominance one, the only real difference being that the phenotype of the heterozygote is different from that of the homozygous dominant in the case of incomplete dominance. Okay, any questions now? Yes, over on the right, yes sir?

STUDENT: I was thinking about inherited diseases and things as you were talking. If there is a genetic disease that isn't very common, are you always the homozygous recessive genotype if you have the disease?

FARNSWORTH: Excellent...excellent question. No, not necessarily. It differs from disease to disease. Some diseases are expressed, or result, if you have the recessive homozygous genotype. Others result if you have the dominant or heterozygous genotype. And the abundance of a gene in a population doesn't have anything to do with whether it is dominant or recessive. You can have a rare dominant gene, or a rare recessive gene. For example, there is a condition humans have sometimes that is called *polydactyly*, in which the person has extra fingers or toes. Polydactyly is extremely rare, yet it is produced by a very rare dominant gene. If a disease is expressed in the homozygous recessive genotype, a heterozygote could carry the disease gene and possibly pass it on to his or her offspring, but not actually have the disease. Such an individual is called a *carrier*. On the other hand, if the disease is expressed by a completely dominant gene, a homozygous dominant individual or a heterozygote will have the disease, but a homozygous recessive one won't.

Okay, that was a good workout. Next time, we'll look at problems in which you consider more than one gene at a time. (*He exits.*)

Day 12

Crosses

Professor Farnsworth enters brushing snowflakes from an unusually early storm off his Mackinaw jacket. On his head, he has a Peruvian snow hat with earflaps and tassels. He stomps the snow from his boots, mumbles under his breath, then begins.

FARNSWORTH: Good morning. It's not, really. I *hate* snow. Ice is nice in a glass, but it's disgusting on the ground. I am cold to the bone, so you're going to have to warm me up with good answers—nothing like a fine answer to generate heat.

Last time, we talked about simple crosses, where you were looking at only one trait. We also looked at a couple of sets of rules, complete and incomplete dominance, that determined what the relationship was between genotypes and phenotypes. Today, we'll take a look at some more complex situations, but if we take the complexities one at a time, they won't be too bad.

The first case I want to consider is the situation in which you want to know what the genotypes and phenotypes of offspring will be in a cross, when you are looking at more than one trait at a time. The simplest of these cases involves two traits, with the genes that determine the traits on different chromosomes. We call this a **dihybrid cross**. The first trait we looked at was werewolfism, which is a completely dominant trait. Let's stick with that one, but also look at another trait. Ummm, let's see, another imaginary inherited one. I'm going to use imaginary traits for a while because the real ones so often have exceptions and complications that it is often a little tough to follow the *main* idea and still be biologically accurate....Got it! Nerdism. Nerdism is completely dominant. If you are a nerd, you are either an *NN* or an *Nn*, and if you are an *nn*, you are cool. By the way, never laugh at nerds. Be nice to them. I was a nerd as a kid, and all my nerdish friends grew up to own their own computer companies and hired the cool people to sweep floors. Now as it happens, the *N* gene is on a different chromosome than the *W* gene for werewolfism, so let's see what happens to these two genes at meiosis, which is a good starting point. (*He places a transparency on the overhead projector.*)

First of all, notice the pair of chromosomes with the *W* gene that have just moved away from the equator of the cell. Let's say that this person is heterozygous for the *W* gene, a *Ww* in other words. Our *W*-bearing pair of chromatids are above the equator, and the *w*-bearing pair are down below. Remember from

175

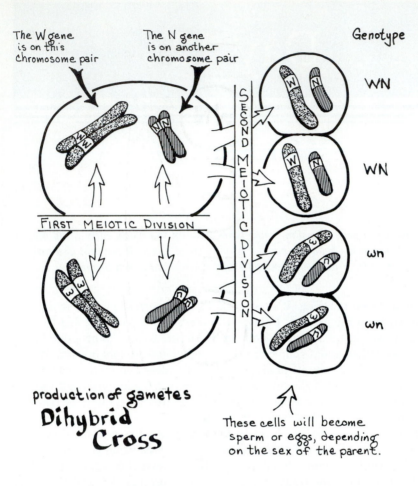

The W gene is on this chromosome pair

The N gene is on another chromosome pair

Genotype

SECOND MEIOTIC DIVISION

WN

WN

wn

wn

FIRST MEIOTIC DIVISION

production of gametes

Dihybrid Cross

These cells will become sperm or eggs, depending on the sex of the parent.

our discussion of meiosis that these positions could just as easily be reversed; the *W* could have been on the other side of the equator as well. Now as it happens, the chromosomes that contain the *N* gene are right next to the *W* ones, but that part doesn't matter—they could have been anywhere along the equator. They are next to each other just for convenience in drawing them. Let's say that the person is heterozygous for nerdism. What kind of phenotype does that make this person then?

STUDENT: A nerdish werewolf. (*The class chuckles.*)

FARNSWORTH: Boggles the mind, doesn't it? I wonder where he'd put his pocket pen protector? Anyway, look at the way I've drawn this. The *N* is above the equator and the *n* is below it. Okay, new drawing here. (*Places another transparency on the projector*) When the homologous chromosomes get separated during the first meiotic division, as you can see, the chromosomes with the *W* and *N* alleles go into the

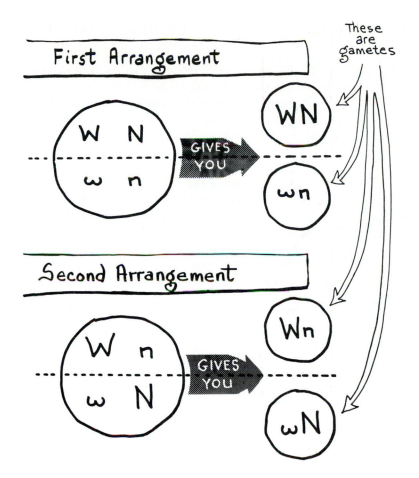

These are gametes

First Arrangement

W N
ω n

GIVES YOU

WN

ωn

Second Arrangement

W n
ω N

GIVES YOU

Wn

ωN

upper cell, and the ones entering the lower cell have the *w* and *n* alleles. The second meiotic division then takes place, and what do you have? Four possibilities for gametes: two of them have a *W* and an *N* allele; the other two have a *w* and an *n* allele. We could call these *WN* and *wn* gametes.

Are *WN* and *wn* the only kinds of gametes we could get from somebody with a *WwNn* genotype? Nope. See, in the first case we looked at, the chromosomes carrying the dominant alleles were on the same side of the cell's equator. No reason for this arrangement; just chance. Equally probable is the situation in which one of the dominant alleles is on one side, and the other dominant is on the opposite side of the equator. If you arrange them this second way, after the second meiotic division you'll have *Wn*- and *wN*-bearing gametes. This gives you a total of four different possible kinds of gametes, *WN*, *Wn*, *wN*, and *wn*, and you're equally likely to produce one as any of the others. Do you begin to see now how meiosis, by scrambling up the alleles on the different chromosomes, can produce a lot of variation in the offspring? Okay, any questions?

STUDENT: Yes, in the example you used, could you get something like a *WW* gamete?

FARNSWORTH: Ah, glad you asked that. No, you can't, but that's a mistake a lot of students make when they don't understand meiosis. Since you have one member of each pair of homologous chromosomes going to each gamete, and each chromosome carries only one allele of a given gene, you can't have a *WW*, because that would mean that the gamete would have two *W* alleles.

Now, having said that, there are certain circumstances when something goes *wrong* with meiosis, and instead of going to separate cells at the first meiotic division, both replicated chromosomes go into one of the offspring cells, leaving the other offspring without that particular chromosome. This failure to split apart is called **nondisjunction**, and it usually has disastrous consequences if one of the defective gametes either fertilizes, or gets fertilized by, an opposite sex gamete. A lot of times the zygote just dies, or develops abnormally.

Going back now to the normal process, I think you can see that you could work out the possible gametes for any number of alleles on different chromosomes. For example, if there were a third gene you were interested in, say the *R* gene for rudeness, you would have eight different gamete possibilities if you started with somebody who was a *WwNnRr*—you might want to work that out as an exercise after class.

So now we know how to figure out what kind of gametes a given genotype can produce in a dihybrid cross. For example, somebody tell me what kind of gametes a heterozygous cool werewolf could produce?

Short pause, then hands up all over the auditorium.

MANY VOICES: Wn and wn.

FARNSWORTH: That's right. A heterozygous cool werewolf would be a *Wwnn*. Now we're ready for the next step: matching up the gametes from different individuals. In principle, it's done the same way we did a monohybrid cross, using a Punnett box.

Let's start with an example. Say we wanted to see what kinds of offspring could be expected in a cross between a *WWNn* and a *WwNn* individual. Both of them would be nerdish werewolves, I guess. Let's say the *WWNn* one is the male.

First, on the vertical axis, you list the kinds of sperm he could produce: *WN* and *Wn*. Then, on the horizontal axis, list the eggs the female could produce: *WN*, *Wn*, *wN*, and *wn*. Draw your box, and you have eight cells. Match up the eggs and sperm, and that gives you the possible genotypes of the offspring. Now, if they had a lot of kids, how many of them would you expect to be cool? That's right, a quarter of them.

As another exercise to do after class, I'd like you to cross a double heterozygote with a double heterozygote, that is, two *WwNn* genotypes. If you've done it right, you'll have 16 cells in your box and you'll see a ratio of 9:3:3:1 in the phenotypes. You'll see only one cool non-werewolf, for example.

Okay, this scrambling of chromosomes at meiosis can produce a lot of new combinations of alleles, but if we go back to meiosis for a second, I'll show you another way variation can get introduced in mating. (*He places a transparency on the overhead projector.*)

This figure shows a pair of replicated chromosomes lined up at the equator during the first meiotic division. Let's say there are only five genes represented on this chromosome, *A*, *B*, *C*, *D*, and *E*. The individual is an *AaBbCcDdee* genotype, okay? One replicated chromosome has the arrangement *AbCDe*, and the other one *aBcde*. In the simplest case, the two replicated chromosomes would part at the first meiotic division, and the individual could produce two kinds of gametes, *AbCDe* and *aBcde*. But there's a complicating factor. The actual physical chromosomes are long and skinny, and like everything else in the cell, they're in constant motion because of molecular movement. So the chromatids are just *whipping* around, and it most often turns out that a chromatid from one homologous chromosome will cross over a chromatid from the other chromosome. The chromatids then some-

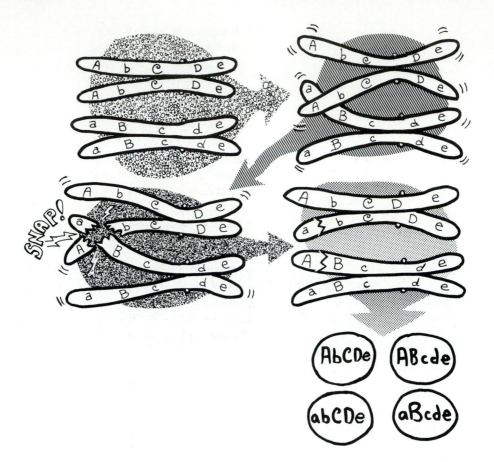

times break at this junction point, or *chiasma,* and the two crossed-over chromatids swap ends. In this case, the break occurs between the *A* and *B* gene loci. After the chromatids have "healed" and the cell has gone through the rest of meiosis, the individual can then produce not just two different kinds of gametes, but four. The break point in the chromatids could occur anywhere along the length of the chromosome, so more than one gene could be exchanged. This process is called *crossing over,* and it is a very important way of producing new variations in sexually reproducing species.

   Speaking of sex, if you have a kid, what are the odds that it will be either a boy or a girl?

   STUDENT: A hundred percent.

   FARNSWORTH: Good; you're starting to listen. What else *could* it be other than a boy or a girl? There's a 100 percent that it will be either one. Slightly different question. If you have a kid, what are the odds that it will be a boy? Well, certain complications notwithstanding, common sense sort of tells you it will be one-half. Let's try to establish it genetically, because if you can do this, you'll be able to do a lot of genetics problems later which involve the same principles of probability.

Sex in humans is determined by a pair of chromosomes called the *sex chromosomes,* the X and the Y. In some senses, they're like a pair of homologous chromosomes because they line up opposite each other at meiosis, but physically, they're different— the X is larger. The information on these chromosomes determines gender, but the chromosomes also carry information about other phenotypic features, for example, some of the characteristics of the blood. In the normal situation, if you have two X chromosomes, you're an XX and you're a female. If you have an X and a Y, you're an XY and you're a male. Okay, now before we do the problem, I have to tell you one thing. Take the female, the XX. It is true she has two X chromosomes, but it will be important to note that one of those X's came from her mother and the other came from her father—just take my word for it for right now—you'll see how it works in a minute. So let's identify the one that came from her father with a little superscript f, like this: X^f. So she's really an X^fX. In the male, his X is a plain X, not an X^f, because his X chromosome came from his mother and his Y came from his father.

Now we're ready to make a Punnett box, but with a little difference here. (*He places a transparency on the projector*; see page 182.) He can produce an X and a Y, and she can produce an X^f and an X; no problem here. But let's throw some numbers in. Take his X. How many of his sperm are going to be X bearers? Half. What are the odds of any one sperm being an X bearer? Again, half, so we write it down. If we ask the same question about his Y's, same thing, half and half, so we write down "½" next to each gamete possibility. We ask the same questions about her. What fraction of all her eggs are X^f bearers? Half. X bearers? Half. Write the numbers down again.

Now, when we match gametes, just as we did before, we ask one more question. Take the upper-left cell. What is the *chance*, the *probability*, that an X^f egg from her will be fertilized by an X sperm from him? A simple rule called the *rule of multiplication* will tell us. It says that the probability of two independent events happening at the same time is the product of the two events happening by themselves. So in this situation, we are saying, "What is the probability that an egg from her will be an X^f *and* that a sperm from him will be an X?" One-half times one-half, or one-quarter, and we write it down in the cell; one-quarter. We ask the same question and do the same thing in each of the other cells.

What do we have now? Four possible genotypes: X^fX, X^fY, XX, and XY. Each of these four genotypes is equally probable, so the probability of a given child's having a specified one of these four genotypes is one quarter. How about phenotypes? Of the four genotypes, two, the X^fX and the XX, are female, and two, the X^fY and the XY, are male. Two out of four is one half, so half the offspring will be male, and half will be female.

A couple of points to observe here. Which gender determines what the sex of the offspring will be?

STUDENT: The male.

FARNSWORTH: That's right. The female has only X chromosomes to contribute, doesn't she? The male, on the other hand, is the only source of Y chromosomes, and you have to have a Y to be a male. Now, it sometimes happens that for a variety

of reasons more of one kind of gamete produced by a person will survive than the other kind. For example, maybe more X's will survive than Y's. So if a man wants to have sons, as is true in many cultures, and his mate produces only daughters, where does the responsibility most likely lie? Figure it out yourselves.

Second point. I made a big deal about identifying these X's from the female as being from either her mother or her father. Here's why. All the genes on a chromosome are physically tied together, or *linked*, so unless there is crossing over or something like that, all the genes on a given chromosome are going to go into a gamete *together* during meiosis. Remember, I said that the sex chromosomes carry not only information for the determination of gender, but also information relating to other things as well. Suppose there is something wrong with one of these genes for other things, a defect of some sort. Let's say that this defect is on the X chromosome. What is the effect of this defect? Let's make one up. Ummm, the effect is that you develop into someone who has terrible cravings for, uh, tuna swirl ice cream. You need it every week or you die.

Okay, let's start with a woman who has one of these tuna genes on one of her X chromosomes. Let's call the normal X chromosome X, and the tuna chromosome X^t. She has one X and one X^t, so she is an XX^t. She can produce both X-bearing and X^t-bearing gametes. Okay, she gets married to a normal man, so let's draw a Punnett box. What kinds of offspring can she produce? Four kinds: an X^tX female, who hates tuna ice cream but can contribute an X^t to the next generation; an XX female, who is normal in every respect; an XY male, who is normal in every way; and an X^tY male, who has to get weekly fixes of tuna ice cream. Why is he afflicted? Doesn't he have only the one recessive X^t allele? Why didn't the female who was X^tX show the effect? Because she had a normal allele on the other X that masked the effect of the X^t allele. Since the Y chromosome doesn't carry *any* tuna alleles, if the man has an X^t, he'll have the fishy condition.

So what we see here is that this tuna swirl gene is not inherited in the simple way that we talked about before. It is *sex-linked*, so that the appearance of the characteristic is dependent on gender. Yes, question?

STUDENT: Yes, in a sex-linked gene situation, could a female ever have the condition?

FARNSWORTH: Good question. Yes, possibly, but it wouldn't be likely. Here's how it could happen. Suppose the daughter in our example, who was a carrier for the X^tX genotype, married a tuna swirl man, an X^tY. Do a quick Punnett box, and you see that they would have X^tX^t, X^tY, X^tX, and XY offspring. One of the boys would be a tuna swirl lover, but so would one of the girls, so there you have your possibility. The reason for the unlikelihood is that many of the real sex-linked defects produce an early death, or severe disability, so males showing the sex-linked character just don't make it to the age of reproduction and, consequently, don't

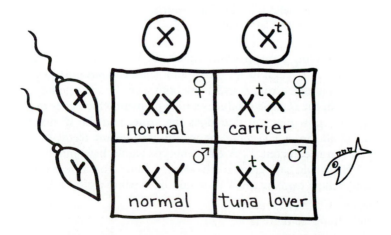

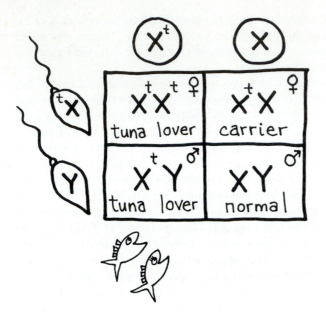

contribute the allele necessary to create a female homozygous for the trait. So, yes, it is biologically possible to have a female showing a recessive sex-linked character, but it is not realistically likely.

Okay, one more topic and we can call it a day. So far, we have been talking about genes for which there are two alleles, a dominant and a recessive, an *A* and an *a*. Well, actually, in nature it is kind of unusual to find only two possible alleles. It is more common is to find three or more. Maybe an *A*, an *a*, and an *a'*. A lot of students get confused here. In this situation, the *A*, *a*, and *a'*, there are three possible alleles, but the genotype of the organism can have only two at a time, one on each member of a pair of homologous chromosomes. So you could be, for example, an *AA*, an *Aa*, or an *aa*, but you could also be an *Aa'*, an *a'a'*, or an *aa'*. A gene that has more than two possible alleles is said to have **multiple alleles**. You can read about real examples in your text, but let's make up one—the principle is the same, no matter what the example is.

Hmmmm. Got it, tooth color. Three possible alleles, *B*, *Y*, and *W*. Both *B* and *Y* are completely dominant in relation to *W*, but *B* and *Y* are **codominant** to each other; that is, the phenotypic effect from *B* and *Y* are both seen. Possible genotypes and phenotypes now? Well, the *B* allele produces a blue pigment which goes to the teeth, the *Y* produces a yellow pigment that does the same, and the *W* doesn't produce any pigment at all. So, a *YY* genotype would have yellow teeth, but so would *YW* because all you need is one *Y*. A *BB* genotype would have blue teeth, and so would a *BW*. A *BY* would have green teeth—lovely, huh?—because it would produce both blue and yellow pigment. A *WW* would have white teeth because it doesn't produce any pigment at all.

Now, you can have some complications in a multiple allele system. One of the alleles could be completely dominant with relation to one of the other alleles, but incompletely dominant with relation to another. The only way really to know how the system works is to do crossbreeding experiments.

So far as doing problems is concerned, you handle multiple alleles like any other simple genetics problem. You want to know the possible offspring between a white-toothed and a heterozygous blue-toothed individual? Well, the white-toother is a *WW*, and the heterozygous blue-toother is a *BW*, so you take it from there.

This was a good day's work. Over the last two lectures, you've had the core of what's called *Mendelian particulate genetics;* "particulate" because the genes behave like discrete particles, rather than some sort of fluid that blends together, as the old-timers thought.

Real genetics is a lot more complicated and has a lot more exceptions than what I've shown you, but if you understand these basic principles, the other stuff will come easy.

Next time, we'll look into the question of predicting how many people will have a particular allele in the next generation, if we know how many have it in this generation. (*He exits.*)

Day 13

Population Genetics

Farnsworth enters wearing a bomber jacket, khaki pants, and aviator-style dark glasses.

FARNSWORTH: Good morning. I will say this about cold mornings—if you fly small planes, you get off the ground a lot faster. It's funny—I took flying lessons because I was always afraid to fly, and I figured that the unknown was a big part of fear, so remove the unknown, and you've removed the fear. The only problem is that I used to *suspect* all the things that could go wrong; now I *know* all the things that can go wrong.

Well, the topic for today is a little bit like flying. With airplanes, there are always trade-offs. If you want to fly faster, you have to give up load-carrying capacity. If you want maneuverability, you have to give up stability. We're going to be talking about a biological area that also involves trade-offs.

Last couple of lectures we've been talking about genetics, but it has always been the genetics of the descendants of a pair of individuals—what kinds of offspring will they have, and what is the chance that their offspring will have a certain combination of alleles. But suppose you wanted to know something about the genetics of *groups* of individuals. Suppose you were a public health planner, and you knew that there was some inherited disease—I'll name just one, say, Huntington's chorea. You need to know how many hospital beds are going to be needed 25, maybe 50, years from now for future victims. You need to know this now because it takes a long time to build a new hospital, to raise the money, and so forth. Now, you know how many people have the disease now—can you figure out how many people will have the disease in the future? Yes, but—and this is an important *but*—your calculation will be an *approximation*, not an exact number. The approximation will be good enough for your purposes, however. It's just like my little airplane; I don't need to know that it will take 8.7657 gallons of gas under a certain kind of condition to get to my destination. "About nine gallons" is good enough because the real conditions probably won't be exactly like the ones that I assumed for a more precise estimate. You always think of science as something that is very precise and exact, and it is a lot of times, but knowing how to approximate can be very, very useful.

Okay, there is this island someplace in the south Pacific, a little tiny island miles from anywhere. Let's call it Hackensack—I don't know why, some names are just

funny. Pawtucket, Rhode Island; Pahrump, Nevada; Cucamonga, California. Anyway, the discoverer of Hackensack was from New Jersey, I guess. On Hackensack, there live five people only; maybe it was a shipwreck. Now, we are interested in a certain gene in this group of people that we just barely mentioned last time, the *R* gene for rudeness. Now rudeness is not really an inherited characteristic, but if it were, I'm gonna say that it would be expressed as a recessive; that is, if you're an *rr*, you're rude, but if you're either an *RR* or an *Rr*, you're polite. Back to the people. There's Charlie, who's an *Rr*, Moonrise who's an *RR*, Heather, who's an *RR* also, Justin, who's an *Rr*, and Eric, who's an *rr*. That makes him—(*pauses for effect*) Eric the Rude! (*Groans from class*) Good, somebody knows Norse history. Now, let me make a little table here showing their genotypes. We have an *Rr*, *RR*, *RR*, *Rr*, and *rr*. We call the total number of genes represented by the genotypes in a population the **gene pool**, so how many genes do we have in this pool? Ten. Notice that we're not talking about the actual, physical genes in the cells of the individuals in the population—that would be a number in the millions. What we're really saying here is that the person with the *Rr* genotype could contribute either an *R* or an *r* to the next generation, and so forth.

So we have 10 genes in this gene pool. How many of those genes are the *R* allelic form?

STUDENT: Six.

FARNSWORTH: That's right, so that means that you have four *r* alleles. We call the fraction of the total number of genes in a gene pool represented by a particular allele the **gene frequency** of that allele. It really ought to be the *allele frequency*, but, hey,

I didn't make up the definition. Okay, on Hackensack, the R allele represents 6/10 of the total. The gene frequency is usually expressed as a decimal fraction, so we would say that the frequency of the R allele is 0.6, and of the r allele is 0.4. The frequency of all the alleles in a pool has to add up to 1.0—that's an important thing to remember when you start doing problems.

Okay, on Hackensack today we have one rude person and a gene frequency for the dominant allele of 0.6 and a frequency for the recessive one of 0.4. Now, these are all young, healthy people, and nature takes its course. They have a whole bunch of kids. In this next generation, what fraction of the kids could be *expected* to be rude, and what would be the gene frequencies in the next generation? One-fifth of the people in this generation are rude. Can we expect about the same fraction in the next generation?

To answer this question, I have to set up another situation for you, this time with two islands. (*He places a transparency on the overhead projector.*) Here on the left you

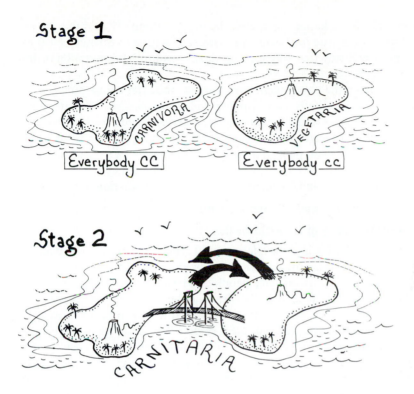

Stage 1

CARNIVORA

VEGETARIA

Everybody CC   Everybody cc

Stage 2

CARNITARIA

have the island of Carnivora. On the right is Vegetaria. The two islands are close, but nobody on either island knows how to swim or has a boat, so the populations are kept entirely separate.

Now, what we are going to look at here is the tendency to eat meat. Like rudeness, eating meat isn't really an inherited tendency, but just for fun we'll say that eating meat is determined by a completely dominant gene C. If you're a CC or a Cc, you eat meat, but if you're a cc, you are a vegetarian. Let's make a couple of starting assumptions about these islands.

Let's say that both populations are exactly the same size, and that everyone on Carnivora is a dominant homozygote. Everybody on Vegetaria eats nuts and fruits. So what is the gene frequency of the C allele on Carnivora?

STUDENT: It would be 1.0.

FARNSWORTH: That's right, so what would be the frequency of the C allele on Vegetaria?

ANOTHER STUDENT: It would be 0.0.

FARNSWORTH: That's good, because everybody is a cc. Okay, on Vegetaria, the frequency of C is 0.0, and c is 1.0. Suppose we came back in three generations; would those frequencies be the same, assuming of course that there weren't any castaways washed up on the beach who had a C allele, or that none of the c alleles got mutated into C's by radiation, or chemicals, or something? Would they be the same? Sure. The same is true for Carnivora. You start with a frequency of 1.0 for the C allele, and it will stay that way forever, again assuming no mutation or immigration.

Now, some guy from the government comes along, and after completing a study says, "Let's build a bridge between Carnivora and Vegetaria." So they do. Now everybody is free to mingle with everybody else. Whereas before we had two separate populations, now we have one big one, whose gene pool is twice as big as the gene pool of either of the small pools that were fused together. We'll call this combined gene pool "Carnitaria." In Carnitaria, what is the frequency of the C allele? Remember, both Carnivora and Vegetaria had the same population size.

There is a long pause, then a student raises her hand.

STUDENT: The frequency of C would be 0.5.

FARNSWORTH: Terrific. Each of the alleles would represent half the new pool. So the day the bridge opens, and the populations mingle, C is 0.5 and c is 0.5. But pretty soon, relationships start up, and people start having families. If we came back and looked at the frequencies of the alleles, and the genotypes and phenotypes made from those alleles, in the next generations, would we still find that Carnitaria was made up of half vegetarians and half meat eaters? Would the gene frequencies be the same? Good questions; now I'll show you how to develop the answers, but I'm going to make some assumptions first. I'm going to assume that there is no preferential mating based on eating habits—a meat eater would just as soon marry a vegetarian as another meat eater, and vice versa. In other words, opposites don't attract, but they don't repel either. The second assumption is that there is no particular survival advantage in Carnitaria in being either a meat eater or a vegetarian; both get along equally well.

There is one final assumption, and that is that we are going to look only at frequencies of alleles and genotypes in the next generation; we're not going to count anybody in the mom-and-dad generation.

To get kids in the next generation, we're going to have to mate people up. Let me make a table here. (*He takes a moment to draw the table on a transparency.*) On the left I've put the kinds of parents you have for each possible mating; on the right, the possible genotypes of their kids. Look at the first one, a CC mated with another CC. What kinds of kids could they have? Do a Punnett box; you'll see that they could produce only CC kids. Now another question. How likely is it that a CC will actually mate with a CC? Remember, people don't make mating choices on the basis of diet. They match up with each other at random. So this guy here in the top of the male column, what's the chance that he's a CC? What's the chance that any male picked at random will be a CC? Well, half the population of Carnitaria are CC, so there's a half chance of any male's being a CC. We write that chance down. Now take the female in a mating. What's the chance that she's a CC? Also half. Now, remembering the law of multiplication, what is the chance that any given male will be a CC *and* that his mate will *also* be a CC? It's the product of the two independent probabilities: $\frac{1}{2} \times \frac{1}{2} = \frac{1}{4}$. We can rephrase the question. Of all the possible mar-

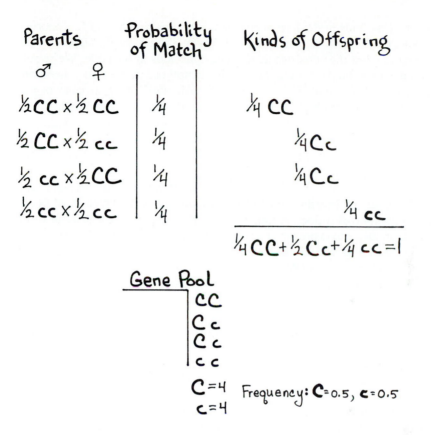

Parents | Probability of Match | Kinds of Offspring

$\frac{1}{2}CC \times \frac{1}{2}CC$ $\frac{1}{4}$ $\frac{1}{4}CC$

$\frac{1}{2}CC \times \frac{1}{2}cc$ $\frac{1}{4}$ $\frac{1}{4}Cc$

$\frac{1}{2}cc \times \frac{1}{2}CC$ $\frac{1}{4}$ $\frac{1}{4}Cc$

$\frac{1}{2}cc \times \frac{1}{2}cc$ $\frac{1}{4}$ $\frac{1}{4}cc$

$$\frac{1}{4}CC + \frac{1}{2}Cc + \frac{1}{4}cc = 1$$

Gene Pool

CC
Cc
Cc
cc

C = 4   Frequency: $C = 0.5$, $c = 0.5$
c = 4

riages in this generation, what fraction of them will have a *CC* male and a *CC* female? Same thing, ½ × ½, or ¼. One-quarter of all the marriages in this generation will be between a pair of *CC*'s.

Now that we've determined this, we're ready to ask about kids. Of all the kids produced in the next generation, what fraction will have a particular genotype, and be the product of a particular match? Well, again, look at this first example. What kinds of kids can be produced by a *CC* × *CC*? We've seen only *CC*'s. So the probability of a *CC* kid from a *CC* × *CC* match is 1.0. So now we get specific. Of all the kids produced in this generation, what fraction of them will be *CC*'s from *CC* × *CC* parents? Well, ¼ of all the matches are *CC* × *CC*, and 1.0 of all their kids will be *CC*'s, so the probability of being a *CC* from a *CC* × *CC* match is ¼ × 1, or ¼. One-quarter of all the kids in the next generation will be *CC*'s from a *CC* × *CC* match.

Move down to the next line, *CC* male, *cc* female. We work it out exactly the same way. What's the probability that he's a *CC*? A half. What's the probability that she's a *cc*? Again a half. What fraction of all the possible matches could be expected to be composed of a *CC* male and a *cc* female? Half times a half, or a quarter. What kinds of kids can they produce? Only *Cc*—do the Punnett if you don't believe me.

So finally, of all the kids in the next generation, what fraction have a *Cc* genotype and are from a *CC* male × *cc* female match? Again, ¼ × 1, or ¼. *Capish?*

I don't have to work the other two matches out; just look at the table. We're now ready to ask the question "What are the possible genotypes in the next generation, and what fraction of the whole is represented by each one?" Well, as you can see, you can have *CC*'s, *Cc*'s, and *cc*'s. One-quarter of all the possible kids will be *CC*'s, and they'll all come from *CC* × *CC* parents. There are two sources for *Cc* kids, and between the two sources, ½ of all the kids in the next generation are going to have *Cc* genotypes. Follow the same reasoning for the *cc*'s, and you find that ¼ of all the kids in the next generation will be *cc*'s.

Add this all up. We have ¼ *CC*, ½ *Cc*, and ¼ *cc*. If we've done this right, and truly considered all the possibilities, this should add up to 1, which is all possibilities, and sure enough it does. Another way of expressing this, instead of using fractions or decimals, is to express it as a ratio. For every 1 *CC*, you will have 2 *Cc*'s, and 1 *cc*. This would be written as a 1:2:1 ratio. Keeping this ratio in mind, we can now ask about the fraction of alleles, the gene frequencies. Refresh my memory, what was the gene frequency of the *C* allele in the mom-and-pop generation?

STUDENT: It was 0.5.

FARNSWORTH: Right. Okay, using that ratio that we just showed, we can make a kind of model gene pool here. For every 1 *CC*, there will be 2 *Cc*'s, and 1 *cc*. That makes eight genes in our model pool; let me count each allele: 1-2-3-4 *C*'s and 1-2-3-4 *c*'s, an equal number, so the frequency of *C* is still 0.5, and of *c* 0.5.

Let's study this for a second. We started in our first generation with half *CC*'s and half *cc*'s, which meant that half the population were vegetarians. The gene frequencies were half and half. Now, the *gene* frequencies are still the same, but only a quarter of the population are vegetarians. It almost looks as if vegetarianism is disappearing. Very mysterious. I wonder what would happen if we let *this* generation mix and match just as we did the last one. Are these ratios going to change? Poll time. How many of you think the fraction of vegetarians is going to go down? Raise hands. (*About one-quarter of the class raise their hands.*) Interesting. Well, let's work it out and see. Now, this is maybe going to *look* complicated, but it is done exactly the way we did it before. (*He places a transparency on the projector.*)

We set this up exactly the way we did in the first generation. Over to the left, columns for the male and female parents. We then have our column for the probability of a match between people with these genotypes. Then columns for the genotypes of the offspring. All right, let's start with a *CC* male and a *CC* female.

Well, in this generation, if we picked a male out at random, what would be the probability that he would be a *CC*? This is really the only tricky bit, right here. In the *last* generation, right after the bridge was built, half the population were *CC*'s, right? But we just worked out that in *this* generation, ¼, or 0.25, of all the people have the *CC* genotype. We use the frequency of the genotypes in *this* generation for our calculation. So the probability of a male's being a *CC* is ¼. The probability that his mate is a *CC* is ¼, too. So what is the probability that a *CC* male will mate with

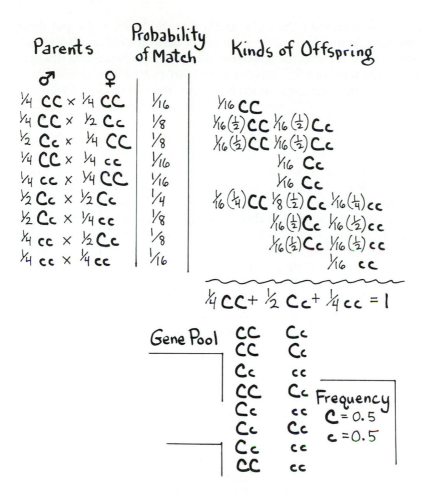

Parents	Probability of Match	Kinds of Offspring
♂ ♀		
¼ CC × ¼ CC	¹⁄₁₆	¹⁄₁₆ CC
¼ CC × ½ Cc	⅛	¹⁄₁₆(½)CC ¹⁄₁₆(½)Cc
½ Cc × ¼ CC	⅛	¹⁄₁₆(½)CC ¹⁄₁₆(½)Cc
¼ CC × ¼ cc	¹⁄₁₆	¹⁄₁₆ Cc
¼ cc × ¼ CC	¹⁄₁₆	¹⁄₁₆ Cc
½ Cc × ½ Cc	¼	¹⁄₁₆(¼)CC ⅛(½)Cc ¹⁄₁₆(¼)cc
½ Cc × ¼ cc	⅛	¹⁄₁₆(½)Cc ¹⁄₁₆(½)cc
¼ cc × ½ Cc	⅛	¹⁄₁₆(½)Cc ¹⁄₁₆(½)cc
¼ cc × ¼ cc	¹⁄₁₆	¹⁄₁₆ cc

$$\tfrac{1}{4}CC + \tfrac{1}{2}Cc + \tfrac{1}{4}cc = 1$$

Gene Pool

CC	Cc
CC	Cc
Cc	cc
CC	Cc
Cc	cc
Cc	Cc
Cc	cc
CC	cc

Frequency
C = 0.5
c = 0.5

a *CC* female? It would be ¼ × ¼, or ¹⁄₁₆, so we put that down in our "Probability of Match" column.

Next step—kids. What kind of kids can they have? Only *CC*'s, so the probability of their having a *CC* is 1.0. Next question: "What fraction of the children in the next generation will be *CC*'s from a match between two *CC* parents?" That would be ¹⁄₁₆ × 1, or ¹⁄₁₆, so we write that down. Now move down to the next match, *CC* male and *Cc* female.

Follow along on the table with me here—I'm not going to mention all the steps in detail. What is the probability that he will be a *CC*? That would be ¼. What is the probability that she will be a *Cc*—be careful here. Half the population of this generation is of the *Cc* genotype, thus the probability that any female will be a *Cc* is ½, so we write that down. Probability of the match would then be ⅛.

What kind of kids could they have? Both *CC* and *Cc*, right? In what proportion? Do a Punnett box if you can't figure it out in your head by now. Turns out that half of the kids could be expected to be *CC*'s, and half would be *Cc*'s. We put those numbers in parentheses right next to the genotype of the kids so we don't forget it. Now, our next question is: What fraction of the kids in the next generation will have the *CC* genotype *and* be from a *CC* father × *Cc* mother match? That would be the probability of the match's occurring, which is ⅛, times the probability that any one kid from the match will be a *CC*, which is ½. That product is ¹⁄₁₆, so we write it down next to the genotype of the kid.

We go down the list, writing the numbers down, then just as we did before, we add up the columns, and look at this! A quarter *CC*, a half *Cc*, and a quarter *cc*. Just like the last generation. Hasn't moved a whisker. Let's check out the model gene pool here; you have 16 possible genotypes from different matches, and lo and behold, it is still ½ *C*'s and ½ *c*'s. There's something going on here.

Okay, to find out what's going on, we're going to use *algebra!* (*Groans*) No, really; see, this just shows you that teachers never lie. Mrs. Frodo, back in the tenth grade, told you algebra would be useful, and it is. All right, we had made an assumption about Carnivora and Vegetaria when we started this. We assumed they had the same-sized population. That meant that when the two populations were combined, the gene frequencies of the *C* and *c* allele would be the same, or 0.5 each. But what if the population of Carnivora were three times bigger than that of Vegetaria, what would be the gene frequency of *C* after the populations combined?

STUDENT: It would be 0.75, or ¾.

FARNSWORTH: That's right. Three out of four people would have the *CC* genotype, so three out of four of all the genes in the population would be the *C* allele. Now, scientists always like to generalize about things, so we're going to give the frequency of the *C* allele, the dominant allele of any gene, a symbol. We're gonna call it the letter p. So if the frequency of the *C* allele is 0.5, we say $p = 0.5$. We're gonna call the frequency of the *c* allele q. We're then going to say $p + q = 1$, which makes sense, because if 1 is the whole quantity, and you don't have anything but the dominant and the recessive allele, then $p + q$ *has* to equal 1.

Then, thanks to the miracle of algebra we can say the following things (*he writes on the board*):

$$p = 1 - q$$
$$q = 1 - p$$

This means that if you know the frequency of *C*, you can quickly get *c*.

Now we're going to see what the relationship is between the frequency of the alleles and the frequency of genotypes. When you mix two populations that are each 100 percent homozygous, one dominant and one recessive, the frequency of the two genotypes will be the same as the frequency of the alleles. So if the frequency of the *C* allele is 0.6, the frequency of the *CC* genotypes will be 0.6, or, using our new terminology, the frequency of the *CC* genotypes will be p. This is exactly the same thing we did the first time, except that instead of using 0.5 or ½ as our frequency, we're using the general term p. Let me make you a new little mix-and-match chart here (*he writes on board*):

$$pCC \times pCC \quad p^2CC$$

pCC × pCC	p²CC	
pCC × qcc		pqCc
qcc × pCC		pqCc
qcc × qcc		q²cc

$$p^2CC \quad + \quad 2pqCc \quad + \quad q^2cc = 1$$

What this equation says in words is that once you've mixed up the two populations, (1) the frequency of the CC genotype is the frequency of the C allele squared, (2) the frequency of the Cc genotype is two times the frequency of the C allele times the frequency of the c allele, and (3) the frequency of the cc genotype is the frequency of the c allele squared.

Now let's try an example or two. If the frequency of the—and let's pick another gene; it doesn't matter which—k allele is 0.6, how many KK individuals will there be on an island that has 1000 people? I'll work this first one out.

First, ask yourself what in terms of p and q are you given? Here, the frequency of k, the recessive allele, is 0.6, so q is 0.6. Then, ask what you need to know. You need to know the frequency of the dominant homozygous individuals. Since this is p^2, you'll have to find p. You know that $p = 1 - q$, and you have q; therefore, $p = 0.4$, and $p^2 = 0.16$. There are 1000 people on the island, so your answer is 0.16×1000, which is 160.

Not so tough, eh? One more. If 64 out of 100 people on an island have the genotype MM, what's the frequency of the m allele? What are you given? Indirectly, you are given p^2. Sixty-four out of one hundred is 64 percent, or 0.64. So p^2 is 0.64. You want to find q, so that means you'll have to find p. You're given p^2, 0.64, and so to get p, which is the square root of p^2, you take the square root of 0.64. Now, if you are an old buzzard like me, you can do that in your head and get an answer of 0.8, but most calculators will let you take a square root. So if you're handicapped in your math, you can use a calculator. Now, if p equals 0.8, that means that q equals 0.2, which is what you needed.

This equation, $p^2 + 2pq + q^2 = 1$, is called the **Hardy-Weinberg equation**, and it is very important to know not only how to work problems with it, but what its biological significance is. What it is really saying is two separate but related things. The first is that if you mix up populations, after one generation of breeding the gene frequencies will reach an equilibrium. This observation leads to the second fact, which is that populations in equilibrium tend to maintain the same gene frequencies over the generations. They don't increase, they don't decrease, they don't wander at random. What is the importance of this observation? Well, let's say that you have some inherited disease, and there are a given number of people in the population who have it in this generation. What Hardy-Weinberg says is that you'll have about as many in the next generation who have it as you have in this generation. That is, of course, unless the disease kills people off before they can reproduce themselves.

You can see that gene frequencies are conservative; that is, unless there is some outside factor, there won't be a change from generation to generation in frequencies. Now you might be saying to yourself, "Whoa there, he made an awful lot of

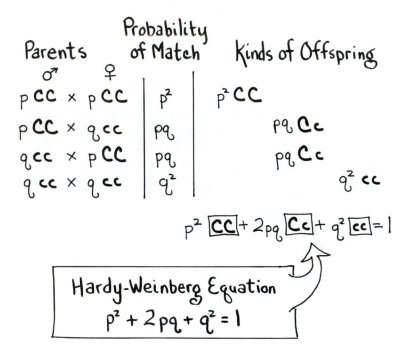

Hardy-Weinberg Equation
$$p^2 + 2pq + q^2 = 1$$

assumptions about this population. No immigration. No emigration. No selective mating. No advantage of one genotype or another. No mutations. That's a whole lotta assumptions. No population in nature is like that!" Well, if you said that, you'd be right, and there's another assumption I didn't mention either. For the Hardy-Weinberg equation to work, the population size has to be pretty big—at least larger than 100 individuals—otherwise you'll get changes in frequencies due to random effects in small populations. With all those assumptions, which probably aren't true for natural populations, what good is the Hardy-Weinberg equation except to give instructors an excuse to put mathematical problems on exams?

The answer is that it provides what is called a *first approximation*. A ballpark figure that gives you someplace to start. If you were analyzing a real island, you would start with a Hardy-Weinberg calculation. Then you would say something like "Well, I know that, for some reason, on the boat coming in from the mainland with new settlers there are more vegetarians than carnivores—so that is going to increase the frequency of *c* alleles. Also, there's so little meat nowadays that a lot of the carnivores die before they reproduce, so that will knock the frequency of *C* down.

So you plug in all the real information that you get from studying a real population, and you can eventually plot a real equilibrium, which might be quite different from the Hardy-Weinberg one.

Now that we're armed with the Hardy-Weinberg equation, the piddling little problem I posed at the beginning of the lecture pales. What about our five island dwellers? Are they destined to be overwhelmed by rude people in future generations? I asked what fraction of the next generation will be little Erics and thus have the *rr* genotype. Since the gene frequency of the *r* allele is 0.4, $q^2 = 0.16$, which is the proportion of brats produced in the next generation. Assuming that the *rr* carriers are not so obnoxious that they can't readily find mates and that while the population is still small a typhoon doesn't eliminate a few crucial members of the gene pool, the gene frequencies on Hackensack will remain fairly constant.

The Hardy-Weinberg equilibrium is at the heart of what is called *population genetics*, the study of the genetics of groups. Once you understand this principle, you're going to be ready to deal with a very controversial topic, which we're going to talk about next time.

I love controversy, so bring your knives and guns next time, and be prepared to argue. I'll tell you in advance what the topic will be so that you can think about it. The topic is...evolution. (*He exits.*)

Day 14

Evolution and
Natural Selection

Professor Farnsworth enters, looking very academic and tweedy, wearing a conservative herringbone sport jacket with elbow patches. Instead of going to the podium, which is his custom, he walks directly out into the class. He stops about a third of the way up one of the aisles and looks around the auditorium, as if he's seeking someone.

FARNSWORTH: I want to play a word association game today. If I point to you and say a word, I want you to give me the first thought that comes into your mind. Don't hesitate, don't think about it, just give me the word—and shout it out, please, so everyone can hear. (*He starts up the stairs of the aisle, but suddenly stops, whirls around, and points to a student wearing jeans and a sweatshirt sitting on the aisle.*)
Food!

STUDENT: I, uh, okay, *pizza.*

FARNSWORTH: Excellent, perfect.

He seems almost frenzied now; he runs up half a dozen steps, backtracks one, and points to the red-haired student he had questioned a couple of lectures ago.

FARNSWORTH: Boyfriend!

STUDENT: (*She's embarrassed.*) Jimmy. No! Gregg. (*The class erupts in laughter.*)

FARNSWORTH: (*Shouting as he runs around the back of the auditorium*) Too bad, Jimmy.

He stops in front of a student wearing a leather jacket.

FARNSWORTH: Evolution!

STUDENT: Uh, Darwin.

FARNSWORTH: (*Still running*) Evolution again!

ANOTHER STUDENT: Uh, monkeys.

FARNSWORTH: (*Continues to run*) Evolution again!

STUDENT: Garbage!

FARNSWORTH: Ah, a man not afraid to express his feelings. Very good. But evolution is both simpler—and more complicated—than any of those answers. Let's get a dictionary definition first. *Evolution* simply means a gradual change in something over time. Car design. Women's fashions. Whatever. The operational phrase is *gradual*. The opposite of evolution is *revolution,* which is a rapid change in something. Nowadays, however, when people say *evolution,* they usually have a more restricted idea. They usually mean what the biologists call ***organic evolution***, which is the gradual change of a species over time. Another way of expressing it, a more technical way, is saying that evolution represents a gradual change in the gene pool and gene frequencies of a population over time.

Over the past 100 years, evolution has been one of the most controversial ideas in western thought. It has been condemned, praised, feared, and promoted. Before you can make an intelligent decision about which of those positions you want to hold—and it is an important decision to make because it affects so many other areas of life—you owe it to yourself to understand just what evolution is.

What I'm going to do today is describe what evolution is, without giving any evidence for it. For evidence, I refer you to your text. This is so we'll at least have a common vocabulary. And the way I'm going to show you what it is is to ask you some questions, and then ask if the answers I propose are reasonable, common-sense, and plausible. Now *plausible* doesn't always mean "true," but one step at a time. Before I begin, however, I have to make a request. Will anyone here in the class today who is not a member of the species *Homo sapiens* please leave the room. (*He looks inquiringly out over the class.*) No one wants to leave? Is it safe to assume then that we are all members of that species?...Okay, let's proceed.

First look at your neighbors. We are all the same species, but are there any differences between you and your neighbors? I think you would agree that there are, true? There *are* differences, and some of them are quite major. So point 1 is that there are differences between individuals in the same species. Sometimes the differences are so huge that it stretches the idea of *species*; for example, a Pekingese and a St. Bernard both belong to the same species, but there are enormous differences between them.

Now, some of these characteristics that are different between individuals—can these characteristics be inherited? We just spent three lectures showing that this was true. So the answer to the second question is "yes, characteristics can be inherited." This is point 2.

Let me ask you this. We know that at least to a certain extent the tendency toward being overweight can be inherited. Suppose, strictly as a thought experiment, I stripped everyone in this room naked, then turned the thermostat down to 50° below zero. Would everyone have an equal chance of survival? Probably not. Who would have a better chance? The overweight ones, because they're better insulated and have more fat reserves to generate heat.

So, let's say we started with 300 people, of whom 30 had an inherited tendency to be overweight. Ten percent. Down goes the temperature. Of the overweight ones, 10 don't make it, but 150 of the slim ones go to the great deep freeze in the sky. So now 20 out of 140 of the survivors, or almost 15 percent, are going to be hereditarily overweight. The frequency of the hypothetical allele for plumpness is increased among the survivors. That leads us to points 3 and 4. Point 3 is that the environment, the surroundings, will have different effects on different phenotypes; point 4 is that some genotypes, the ones that produce the favored phenotypes, will be selected by the environment, and genotypes which produce unfavorable phenotypes will be selected against. An important point here. Not every overweight person lives; not every slim one dies. The *odds* are just better if you're overweight. It is *chance* that determines which *individual* makes it or not. Another important point. Suppose I had the same population I had to start with, 300 people, 30 of them overweight, but this time I didn't put them in the freezer; this time I cranked the heat way up. Who would do better? The slimmer ones—less stress on the heart due to heat shock. So, depending on what the environment is like, one phenotype will be favored over another. Change the environment, and you change the advantage.

Okay, have I said anything so far that sounds farfetched or implausible? Anything sound unreasonable? No? Okay, let me recap for a second. I'm saying first that there are variations within the population of a species. Second, that some of those variations can be inherited. Third, that the environment acts differently on different phenotypes. And fourth, the differential action of the environment can result in a change in the fraction of individuals who have a given genotype from generation to generation.

I know what some of you are thinking. You're thinking, "Well, that can account for a reshuffling of the proportions of different phenotypes in a population from generation to generation. But you're not really going to *change* things. If you start with a population of different breeds of dogs, the environment might favor certain breeds, but you're not all of a sudden going to find that the population of dogs has changed into a population of cats. To do that, you'd have to have brand new alleles in the population, not just new combinations of old ones." Well, if you said that, you'd be right, but there *is* a source of new alleles in a population, and that source is **mutation**. A mutation is just a change in the sequence of base pairs of DNA. The result of this change is the creation of a new allele. What produces the change? Random environmental factors like cosmic rays, rock radiation, and ultraviolet radiation from the sun.

If I said the word *mutant* to people, mostly I'd get a response that was negative. Like "Killer Mutant Radishes from Mars." But a mutation is just a change—it can

KILLER MUTANT
RADISH FROM MARS

have a beneficial, inconsequential, or negative effect, depending on what it does and the circumstances when it occurs. For example, to a population of organisms that is very well adapted to the environmental situation it lives in—like sharks living in the open ocean—any real change caused by mutation is likely to have a negative effect. On the other hand, if the climate is getting colder, thicker fur on mice would be advantageous. If a flock of birds gets blown by a storm to an island that has a harder kind of seeds from what they're used to eating, well, maybe a mutation that produces a thicker beak would be good.

It is very important to realize that there is no idea of deliberate attempts at progress or a purposeful striving toward some goal of perfect adaptation in evolution. Chance drives the process. Let me give you an example. I have here an Indianapolis 500 race car, right on stage (*gestures toward imaginary car*). The race is tomorrow, and the mechanics have been working on it for weeks and weeks, tuning it to perfection for the conditions at the Indy track. Every time they make an adjustment, they take it out on the track, always adjusting for more speed. Now it is 3 A.M., and nobody is in the pits. Suddenly, over the pit fence comes an escaped lunatic called Screwdriver Sam. They call him this because he has an obsessive desire to turn screws. He sees our car, his eyes light up, he whips out his screwdriver, and he comes over to the car and closes his eyes. He feels around with the tip of his screwdriver until he finds a screw that turns. With trembling hands, he turns it a quarter of a turn to the right.

Now, after he does this—this chance finding of a screw and then the chance turning it in a particular direction—what do you think the probability is that this change, this *mutation*, in the adjustment of the car is going to result in an *improve-*

ment in its performance ? Slim to none, right? The car already was about as close to perfect as it could be.

Okay, the Indy race is over; it is time to ship the car to Phoenix for the Phoenix 500. Now, the Phoenix race is a lot different from the Indy 500—the track is shorter, the paving is different, the altitude is different. We know we're going to have to make a lot of adjustments. So we ship the car down, and now it is the night that it arrives. We haven't had a chance to do our testing and adjusting yet. It's 3 A.M., nobody around, and who should come over the fence but Screwdriver Sam! He rubs his hands together, pulls out the screwdriver, closes his eyes, finds a screw, and turns it a quarter of a turn to the left. Now, what is the chance that this random turning is going to result in an improvement of performance? Much better than at Indy, right? We know we're going to have to make a lot of adjustments for the new track anyway; maybe Sam accidentally stumbled on just the adjustment that we need for the new situation. Oh, sure, it could still possibly have a negative effect, but the *probability* that the effect will be good is better when the environment is new.

There's a very important point in this. Sam didn't have a *purpose* in his screw turning. He was doing it just for the hell of it. But the *effect* of his purposeless act might *look as if* he had a purpose in mind. "Oh, wow, that guy is a natural tuner! How did he know that turning the injector needle a quarter left was gonna be just right for Phoenix?!" So when you look at the evolutionary record of plants and animals, and you see relatively simple organisms first, and they get more and more complex, there's a temptation to say, "Clearly, evolution is *progressive* and is going in the direction of greater perfection. Somebody must have designed it that way." Maybe so, but not necessarily so. As we have seen, a chance event occurring at the right time and place can *look as though* it was done on purpose.

So going back to mutations—they can have a beneficial, neutral, or negative effect, depending on how well-adapted the organism already is. What they *do* do is provide the raw material upon which the environment can act—the environment decides what the effect of the mutation is.

The whole concept of the environment's acting to select out certain phenotypes, thereby increasing the probability that those phenotypes will send their characteristics (through the vehicle of their genes) into the next generation, was put together in the middle of the nineteenth century by a man named Charles Darwin. Actually another man, named Wallace, had the same idea at the same time, but unfortunately he's not remembered much today. Darwin called the action of the environment upon phenotypes **natural selection**, to differentiate it from artificial selection, which is what animal breeders practice. There is a huge difference between artificial and natural selection. Artificial selection is done with a purpose, for example, to breed a bigger pig. With natural selection, you might *end up* with a bigger wild pig, if being big is advantageous, but it is not necessary to suppose that there was any design or purpose to the process. This may seem like a fine point, but it has been at the heart of the principal controversy about evolution for the last 100 years, and we might as well air it right now.

When you look around at all the wonderful things there are in the world, how complex they are, it is very tempting to say, "Clearly these things could not have happened by chance; it is too improbable. Somebody, or something, must have designed and created them." That is a perfectly reasonable observation. But as we have seen, *reasonable* does not *necessarily* mean "true." There might be another, equally plausible, explanation.

For example, you might say, "Look at the improbability of arriving at something as beautifully engineered as an eye, or a bird's wing, simply by chance." A very astute observation, but as I will show you, there is chance, and then there is *chance*.

He steps over to a table covered by a sheet. He sweeps the sheet away to reveal 100 glass jars, each containing a quantity of sand.

FARNSWORTH: All right, I have exactly 100 jars of sand here. Ninety-nine of them contain exactly 1 pound of sand. One contains 1 pound and 1 ounce of sand. I don't know where the heavy one is on the table. What is the maximum number of weighings I would have to do to be *100 percent* sure of finding the heavier jar? (*The class buzzes as they try to figure the problem out.*)

STUDENT: The maximum would be 99 because the heavy one might be the last one you weighed.

FARNSWORTH: Okay, hold that thought. Now, don't try to calculate it, just gut feeling, how likely do you think it would be that I'd find it in 10 or less weighings? Pretty unlikely, huh? *Very* unlikely, huh?

Well, in actuality, I can be 100 percent sure of finding it in no more than six weighings, if I use a double balance. What I do is divide the group of jars in half, one group now containing 50 jars, the other also containing 50. I weigh the two groups. One is heavier than the other. So I *know* my heavy jar is in that group. I throw out all the jars in the other group. Now I take my 50 and divide them into two groups of 25, and do the same thing again. One of the groups contains the heavy one. Chuck the other group. I now have 25 jars. I weigh one group of 12 against another group of 12, with 1 left over. If the two groups weigh the same, my lone jar is the heavy one and I've found it after only three weighings. If they're different, toss out the lone one and the light group. Weigh six and six, toss out the light group. Weigh three and three, dump the lights, and weigh two of the three remaining. If they're the same, the odd one is the heavy one; if they're different, you've also found the heavy one. Now, there's a little change I could make in the procedure that would let me find the heavy one in only *five* weighings. See if you can figure it out by the end of the lecture.

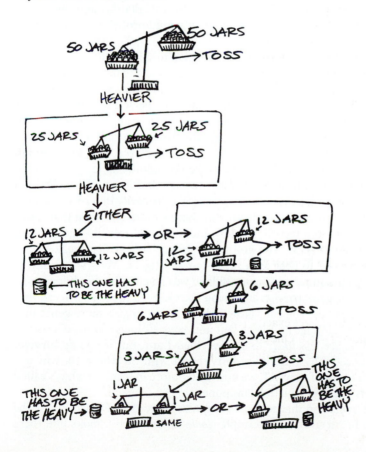

What is the point of this? Well, the heavy jar is analogous to a successful structure— a simple eye, a simple wing, what have you. This simple structure is buried in millions of structures that will turn out to be not so successful—wings are great on birds, but they're not so good on elephants or alligators. Natural selection doesn't have to go carefully through every conceivable structure and pick out the one successful one. All that has to happen is that the structure is included in the group that makes it into the *next* generation, just like our group of jars. With this next generation, selection doesn't have to pick from millions of structures, just hundreds of thousands. Each generation reduces the number of structures natural selection has to pick from.

What this amounts to is that the probability that very complex structures, like eyes or wings, arose by chance is much greater than it might seem to be.

Now, before somebody beats me to the draw, I'll ask the question myself— "Does the theory of organic evolution prove that God didn't create all the animals and plants?" Do you understand now why evolution is so controversial and emotional a subject? Well, the answer is "No, the theory of evolution does not prove that God didn't create the animals and plants. What it *does* do is offer an alternative hypothesis. It is not *necessary* to suppose that God created the animals and plants. But it is not impossible that God did do it. Maybe God works through the mechanism of evolution. Using the methods of science, it would probably be impossible to prove that God *didn't* do it.

So what we have now is that the person who wants to believe that God created the animals and plants is going to have the nagging annoyance of having to deal with a plausible alternative explanation that doesn't depend on God for the existence of the different species of organisms. On the other hand, the person who believes that God didn't have anything to do with it is going to have the nagging annoyance of not being able to demonstrate conclusively, in a scientific way, that it is *impossible* for God to be involved. So now *everybody* can be uncomfortable, but, hey, my job is to give you something to think about, not make you comfortable.

All right, enough of the philosophical. Back to nuts and bolts. When Darwin developed his idea of natural selection, he was thinking primarily in terms of simple survival. You can't have kids if you're dead. But there's more to it than that. Suppose I'm big and strong and a terrific fighter, and so are all my siblings, each of whom I share half my genes with. You and your siblings are smaller, weaker, and not nearly so well equipped to deal with saber-toothed tigers. My siblings and I, however, are lousy parents and abandon our kids so we can go off and fight dragons. You and your siblings are great parents and bring your kids all the food you can find and fight the tigers off as best you can. Now, who is going to pass more genes into the next generation, my siblings and I, or you and your siblings? Well, it's sort of a toss-up, isn't it? You are less *likely* to survive than I am, but if you *do* survive, you will probably successfully raise more kids than I will. So where Darwin was thinking primarily in terms of survival, we now think more in terms of what's called *fitness*, which reflects the reproductive component.

Another factor in evolution that we now recognize more than Darwin did is the *direct* role of chance in survival. For example (*looks around room*), looks like there are about four real redheads here in the room. So the gene frequency for red hair in a class of 300 or so has to be low, right? Well, you don't know it, but this auditorium was built by an inept contractor, and at any time, especially if someone were to fall asleep and snore at the resonant frequency of the building, the roof could cave in. Well, depending on where you were sitting, you'd be killed or you wouldn't be, and whether you were killed wouldn't really depend on how big or strong you were, but just the luck of the draw in where you were sitting. Now suppose, just by chance, the redheads were sitting in the right places. There are only 10 survivors, and 4 of them are redheads. What would be the frequency of the red-haired gene among the survivors? Very high. And would this be because of any selective advantage of red hair? No. Just chance. We're now discovering that chance probably plays a more important direct role in evolution than we thought before.

Well, Darwin missed the importance of chance, but he did pick up on another very important factor in evolution, something that very strongly shapes the character of a gene pool. All right, you're all college students here. If you're deciding whether to go out with someone, do you let chance select a date for you? No. Are some individuals more, what shall we call it, *attractive* than others? All right, the women in the class—which is more attractive, a guy who is a tenth-degree black belt in karate and an expert in both sniper rifles and pistols, who can butcher a cow with his bare hands, and who can go without food or water for 30 days, *or* a guy who has a real nice house, more of an estate, actually, drives a Ferrari, is wildly generous with his girlfriends, and talks about how nice it will be someday, after all the trips around the world, to settle down, marry, and have kids? Is it necessary to raise

hands? Okay, the point is that there is not a direct, one-to-one correlation between characteristics important in survival and characteristics advantageous in securing a mate. Darwin called the process of selecting a mate *sexual selection*, and he pointed out that in some cases, characteristics useful in securing a mate might be disadvantageous to survival. For example, male birds are often brightly colored—the brighter you are, the more intimidating you are to other males you are competing with for mates. Unfortunately, the brighter you are, the more conspicuous you are to predators, so you can't afford to be *too* bright—you'd be killer-attractive to the opposite sex, but a hawk would pick you off before you made your moves.

There's a lot of debate among biologists about how sexual selection works. Do females select qualities of size, color, or voice directly from a pool of available males, or are these qualities important in one male's intimidating another male, leaving the "victor" to collect the female at his leisure? It seems to depend on the species.

All right, let me sort of sum up where we are before we move on. Because of genetic factors like recombination and meiosis, there is natural variation in a population. This variation can be inherited. Not all variations are equally likely to survive. Not all organisms carrying these variations are equally likely to find a mate. Not all variants are equally likely to raise young successfully. The more successful variants increase the frequency of their alleles in the next generation. Chance plays both a direct and an indirect role in evolutionary change. Okay, that's a pretty good synopsis. Let me ask you a question. About how many different kinds of animals and plants are there in the world? Rhetorical question—nobody really knows, but it is in the millions. How did so many species come to be?

First of all, what is a species? Well, for over 2000 years, biologists have been sitting around over coffee and arguing about what a species is. You would have thought they would have hashed this out at least a couple of centuries ago, but they haven't, and I'm not going to be able to resolve the matter today. There is one really nice, elegant definition that I sort of like, but there are a couple of nasty, ugly exceptions, just like the viruses when you're trying to define *living thing*. A *species* is a population of animals or plants which actually or potentially breed with each other, but not with members of other populations.

Okay, these millions of species—how does it happen that we have so many? Well, species change basically in one of two ways. To demonstrate the first way, let's say we have some fictitious animals, call them *grelbers*. Grelbers are desert animals and have lived in this particular desert for about, oh, 10 million years. However, 10 million years is a long time, and the earth's climate sometimes changes within such long periods of time. So it is starting to rain more and more on the desert these days. Well, there have always been a few grelbers that have kind of a waxy skin. Didn't either help or hurt. Now, with more rain, your ordinary grelber tends to get waterlogged and develop these horrible fungus diseases, but waxy grelbers seem to do okay. Over time, the fraction of grelbers having waxy skin increases. So, we come back in a million years, and most of the grelbers have waxy skin, but now we can't really call them grelbers anymore, because grelbers don't have waxy skin. Instead we call the new guys *waxy grelbers*. So the gene frequencies of the species have changed over time. Now, clearly, you don't have it that one day you have grelbers and

next you have waxy grelbers. It is gradual, and it takes many, many generations in most cases. This kind of gradual change in a gene pool is called *phyletic evolution* and represents a situation where a new species generally replaces an old one.

The other way you get change requires a little more explanation. Here's an island, a big island; it is cold in the north, hot in the south. (*He places a transparency on the projector.*) You have these furry animals called *ruffers* that live all over the island. Clearly, it would be advantageous for ruffers in the north to have thick fur and ruffers in the south to have thin fur, but the thing is, ruffers are mobile. So a thick-furred ruffer up north could wander down south and mate with a thin-furred ruffer; some of their off-spring would have thick fur, and some thin fur. The same thing could happen with a thin-furred ruffer from the south; it could move north and mate, and selection could never really change the character of the gene pool as a whole.

But then somebody in the Strategic Air Command gets his instructions mixed up, and instead of sending the stealth bomber to Russia, he sends it to *Ruffia*, and it blows a groove clean across the middle of the island, running east and west. The groove instantly fills up with seawater, and since ruffers can't swim, the population is split in

half. *Now* natural selection can start to work in favor of thick-furred ruffers in the north island and thin-furred ruffers in the south. So the gene frequencies in the two separated populations gradually start to change. Thin-furred ruffers get scarcer and scarcer in the north, and thick-furred ruffers get scarcer in the south. This goes on for a couple of million years. Oh, every once in a while, a thin-furred ruffer will get picked up by a typhoon and blown up to the north island, and if it doesn't freeze, it'll still be able to mate with the thick-furred ruffers, but after a couple more million years, a funny thing starts to happen. Because there are so few thin-furred ruffers up north, they start to look, well, *freaky* to the thick-furred ruffers. Pretty soon, the occasional wind-borne, thin-furred ruffer that lands up north goes to bed alone because who wants to mate with something that looks like *that*? Think that's farfetched? Humans share about 97 percent of their genes with chimpanzees, but I'll bet there aren't more than a half-dozen very desperate people in this class who would take me up on it if I said I could get you a date with CoCo the Chimp. Where was I? Okay, now our populations are so different that individual members of one population will not breed with members of the other.

Ring a bell? You got it; that's the species definition, and so we have created two new species, by a process of *speciation*, where we had one before.

A couple of technical points about speciation. You have to have a *heterogeneous environment,* one that is different in some way from one end of it to the other. You don't ordinarily get speciation, even in a heterogeneous environment, because normally there is *gene flow*. However, if for some reason there is *geographic isolation* between parts of the population, natural selection can act differentially. Pretty soon you'll see the development of *reproductive isolating mechanisms* that will prevent members of the two newly evolved species from mating, even if they can now come in physical contact owing to the removal of a long-standing barrier. At one time, we and the chimps had a common ancestor, there was some kind of barrier formed, maybe the raising of a mountain range or something, and the ancestral population split apart. Nowadays, even if a human and a chimp are in the same room, no matter how nice the chimp is, very, very few humans are going to want to mate with it, and I'd be willing to put big bucks that *nobody* is going to want to mate with a gorilla—even if gorillas *are* supposed to be very gentle. Don't get me wrong; I'm sure the feeling would be mutual.

All right, we are now at the end of a long chain that started with DNA structure and replication, went to mitosis and meiosis, then to Mendelian genetics, population genetics, and now evolution. Although we won't be talking directly about evolution any-

more, it is always there. If you want to understand biology—life—you have to understand the concept of evolution. It is the one thing that ties everything together. Now,
I realize that some of you might be a little upset by some of these ideas, because they
maybe don't square with what you've been taught or believe. And the better the student you are and the more you like to think about things, the more you might be upset.
I understand that, and I am very sympathetic. I went through something similar myself. Well, for those of you so affected, may I make a suggestion? More lecturing from
me up here isn't going to help, so if you're interested, right after lecture why don't we
go over to the Union and grab some coffee and what they call doughnuts—doughnuts
by Goodyear—then come back to my lab and talk about it. I've got a couple of hours
before a flight, and that would be a productive use of time. And for everybody, next
lecture we will be talking about *weird* things; I'll have to make a graphic and disgusting
demonstration of something, but before you kill me, just remember that things are not
always what they seem. Oh! Have you figured out how to find the heavy jar in no more
than five weighings? You divide the jars into *three* groups at the first weighing—33, 33,
and 34. (*He starts to put away his jar demonstration as about 20 students come up to the
podium, all talking at once. It is evident that there will be some invigorating discussion back
at the lab.*)

Day 15

Feedback, Control, and Entropy

Professor Farnsworth enters wearing gray flannel slacks and a white lab coat. In one hand, he holds two paper bags, each big enough to carry a half-gallon of milk. In the other, he carries a small laboratory cage, in which are two live laboratory mice. The mice belong to the C_3H strain and are known as "popcorn" mice to lab workers because of their ability to jump to great heights (for a mouse) and land easily on their feet, like a cat. He goes to the podium, opens up the tops of the two bags, and places them side by side on a shelf that is on the front of the podium. He puts the cage, made of clear plastic, next to the two bags so the students can see the mice. The podium is moved back from its normal position near the front of the stage, so that none of the students can see the back of the podium, and therefore do not know that there is a second, open-topped, empty cage positioned on top of the podium, but below and to the rear of the bag on the left. They also do not know that in the left-hand bag there is a sponge about the size and shape of a mouse. Inside the bag is a false bottom, below which is the sponge. With this arrangement, the bag could be turned open end down, and the sponge wouldn't fall out.

FARNSWORTH: Good morning. This is supposed to be a science class, right? Well, let's have a scientific demonstration. But before I begin, I want to remind you again that appearances can be deceiving, and to tell you that as a college student I picked up extra money at parties as the Amazing Rudolpho. Rudolpho could saw people in half, drive swords through their bodies, and shoot them, and, amazingly, they would never get hurt. Assuming everything worked, of course. All right, over here, I have the two subjects of the demonstration (*pulls the top off the cage, reaches in, and lets one of the mice climb into his hand*), and as you can see, they're healthy and vigorous. They've been in training for months for this, haven't you boys? They're ready to go. Now, this is going to be a demonstration of change of state and the consequences of that change. You'll remember, of course, the beginning of the semester, when we talked about the state of what we called "life" at the time. Now these two mice are as similar as I could get. Same strain; they're even brothers from the same litter, aren't you boys?

All right, I think we're ready to go. I'm gonna take the first "mouseonaut" here (*reaches into the cage and pulls one of the mice out by the tail, which is the standard and*

SPONGE
"MOUSE"

FALSE
BOTTOM

approved way to handle mice) and place him in the first "experimental chamber." (*He drops the first mouse into the right-hand paper bag.*) Now we do the same for the second mouse. Have a good trip!

From the class it looks as though he's dropped the second mouse into the left-hand bag, but actually he's dropped it behind the bag, into the open cage hidden by the front of the podium. This is a standard magician's technique, and the illusion is totally convincing because the eyes' depth perception simply doesn't work at the distances found on stage. The famous Chinese linking rings trick makes use of this principle.

   FARNSWORTH: Okay, both mouseonauts are in their chambers. (*He bends over and talks into bags.*) Belt up, boys! And now we can seal up the chambers and take them over to the experimental area.

He rolls the tops of the two bags down and starts to walk over to a small table which is to the left of the podium. As he walks, he gently shakes first the right, then the left bag. Students in the first couple of rows can hear something soft and loose rattling around in both bags. He places both bags on the table on their sides.

FARNSWORTH: Two experimental subjects, two identical chambers. Every experiment has to have a control, so I am going to administer the experimental treatment to the chamber on the left and leave the right one unmanipulated. Now, I get my experimental apparatus.

He reaches under the table and picks up a paddle that looks something like a canoe paddle with the bulk of the handle sawed off. The wide part of the paddle has a slot cut in it widthwise, so that if it hits something, it makes a tremendously loud craa-c-k. *He hefts it, gets a grip on it, swings back, and hits the top of the table, next to the left-hand bag. The sound is startling.*

FARNSWORTH: Ah, it's working well today. You wouldn't believe the years of effort it took to develop this. Well, ready as I'll ever be. Ready, boys? (*He looks back and forth at the two bags.*) Guess so. I will now attempt to change the state of the left mouseonaut.

He gets a double-handed grip on the paddle, swings back, and smashes it with all his strength on top of the left-hand bag. The class gasps as it hits with a resounding crack.

Professor Farnsworth steps back, and the students can see what appears to be the shape of a mouse molded into the bag. He puts the paddle under the table and speaks.

FARNSWORTH: Voilà! I have now, in all probability, changed the state of this mouseonaut. (*As he speaks, maintaining a continuous line of conversation, he picks up both bags and unrolls their tops, so that they're both now open.*) Let me check. (*He peers into both bags. As he looks into the left bag, he grimaces.*) Oh, yuck! Definitely successful. But now we want to consider the implications of this change of state.

Again, maintaining continuous conversation, what the magicians call "patter," he picks up both bags and moves over to the podium. There is a low shelf in the podium, hidden from the class, that normally is used by speakers to store books. Professor Farnsworth puts both bags, open end up, on this shelf. As he speaks, he puts his hands on the podium, but behind the front edge. This would be a normal position when speaking, so it is not conspicuous, but with his hidden left hand, he is able to pull the hidden cage containing the left-hand mouse to the rear of the podium. There is a trapdoor in the bottom of the cage, so that when he pulls the cage back over the edge of the podium, the bottom falls away and the mouse falls into the left-hand bag. This action takes much less time to do than to describe—a couple of seconds is it. Because the class is still a little shocked by what they've just seen, they don't notice the slight shift of body weight necessary to perform the drop.

FARNSWORTH: Both mice were exactly alike at the start of the experiment, but now there is a difference, as I'm sure you would all agree. But what is that difference? It is one of *organization,* or *orderliness.* Before the experiment, we could predict rather precisely where the left-hand mouse's liver was in relation to its spleen. Now we are not so sure—it might be anywhere. (*Does a mock shiver*) The right-hand mouse presumably is about as orderly as it was before we started the experiment. Now, however, we want to ask the question—in the left-hand mouse, what are the long-term effects of this change of state, and consequent change in amount of organization? Well, here, let me take the chambers over where we can see them a little better. (*He takes the two bags, now both open at the top, to a second table to the right of the podium.*) Now, we admit that there is a difference in orderliness between the two mice in the bag now. But how about three days from now? Would the difference be greater, the same, or less? Surely, it would be greater. The left-hand mouse would become more and more disorganized, and would probably smell worse, with the passage of each day. But what if I left these two chambers here for a month— just went away and left them here, untouched. What would happen? (*The class starts to laugh as they get the implications.*) If I just left the right-hand mouse in here for a month, it too would start to get disorganized, and after, say, three months, if I came back, what would be the state of disorder of the two mice? Of course, the right-hand mouse would catch up to the left-hand mouse, and all we would find if we looked in both bags would be some crumbling bones.

But is this inevitable? Could I maintain the order of the mice? (*He reaches into his lab coat pocket and pulls out some mouse food pellets.*) I have here something that is usually called "food." Now I am going to drop an ounce or so of this food into each of the bags. (*The students laugh again at the ridiculous picture of the apparently smooshed mouse needing food.*) Something is different now. Is the right-hand mouse going to become disorganized and join the left-hand mouse in its disorder? No. But how about the left-hand mouse? Because we have given it some food, will that slow the rate at which it becomes disorderly? No. But why not? The answer has to do

with something called *entropy*, and entropy has to do with what we're going to talk about today, which is the control of biological processes. But before we start, I think we'd better check the *real* condition of our two mouseonauts.

He picks up both bags and walks back to the podium. He turns the right-hand bag upside down and shakes the first mouse into the empty cage that he brought into the auditorium. He then does the same thing with the left-hand bag. Naturally the second mouse, perfectly intact, pops out of the bag, but the sponge doesn't, because it's under the false bottom.

FARNSWORTH: Ta da! I want you to meet Bruce and Dick, none the worse for wear. Bruce and Dick have been doing this gig with me for three years. Dick is the left-hand mouse, and he's been trained to roll up into the neck of the bag exactly seven seconds after I put it down. Nice job, boys.

Let's get back to entropy. **Entropy** is a measure of *dis*orderliness. The more disorderly a thing is, the more entropy it has. For example, there is a lot of entropy in my office, if you've ever been in it. Now, there is a principle of physics having to do with entropy that we will have to get familiar with, because it has a lot to do with biology, living systems usually being considered very highly organized. That principle, called the **second law of thermodynamics**, says that the entropy, or disorder of a system, tends to increase over time. This really makes sense. For example, I restore old cars, and I have a beautiful '57 Ford Fairlane 500 convertible. It is absolutely immaculate. The paint is perfect, the engine is perfect, everything is perfect. It is very organized, in the sense that I *know* where every molecule is in relation to other molecules—where every part is in relation to other parts.

That's the way it is today. Now suppose I take it down to South _____(*a very bad neighborhood known to most of the students*) and leave it parked overnight on the street. When I come back tomorrow to pick it up, would it be as orderly and pre-dictable as when I left it? Almost certainly not. I don't know where the wheels would be, I don't know where the radio would be, I might not even know where the seats would be. There would be scratches and dings in totally unpredictable places. The car would be less organized tomorrow than it is today. It would have more entropy.

But that's an extreme case. Suppose I kept it at home, but just left it parked in my driveway? In a month, would it be as orderly as it is today? No. Every time acid rain fell, some of the paint and chrome would get eaten away. The wind could blow gravel and scratch the paint. So if I just *leave it,* sooner or later, but inevitably, it will completely rust apart and become totally disorganized—so disorganized, as a matter of fact, that it will no longer exist as a system. It'll just be iron oxide dust.

That's a hell of a dismal thought. Can I do anything to prevent this from hap-pening? Well, maybe I can't prevent it indefinitely, but can I slow down the rate at which it picks up entropy? Sure. I could build a garage and put it in the garage. I could wax it. I could cover it with plastic and pump inert nitrogen gas into it. Now, what do all these actions have in common?

STUDENT: They all require energy.

FARNSWORTH: Sure. It requires energy to build the garage, and to put on the wax. That leads me to a generalization. If you can pump energy into a system, you can slow down the rate at which it picks up entropy—for example, if I give it a new paint job. And *that* observation is why the two mice—the live one and the dead one—are different. The live one can pump energy—food—into its system and slow the rate at which it becomes disorderly. The dead one can't incorporate energy into its system, so it just follows the second law of thermodynamics and becomes more disorganized over time.

So the ability to input energy is critical to the maintenance of life. But can I do it just any old way? Could I just heat the mouse up in a pot? No—it has to be a *controlled* input of energy. The right amount and kind of energy at the right time to drive the process that is maintaining orderliness. So, after this long introduction, we come now to what this lecture is about—control of biological processes.

(*He walks down into the class, stops by a student, and addresses her.*) Tell me, ma'am, at this very instant, are you thinking about your heartbeat—when you're going to contract your right atrium and how many milliseconds after that you're going to contract your left atrium?

THE STUDENT: Uh, no.

FARNSWORTH: (*Mock horror, with eyebrows raised*) No! But think how important that is—one mistake, and bingo! We call the meat wagon. Why, I would think you would be spending every waking hour thinking about your body processes—when you're going to squirt out a little chymotrypsin for digestion, when you're going to replace the acetylcholine at your nerve endings. And if under romantic circum-stances you were to start feeling "amorous"—think how long *that* checklist of things to do would be! So how come you can just sit here and not think about it?

THE STUDENT: Well, all that stuff goes on, you know, automatically.

FARNSWORTH: Precisely! But it doesn't "just happen." There are some exquisitely honed mechanisms that provide this control, and we're going to learn what they are today. First, I have a question for the car enthusiasts here. Suppose I had two gasoline engines here—same basic design and size. One of them is going to be used in an automobile, and the other is going to be mounted on a concrete pier in an air-conditioned room and used to run an emergency electric generator. Which of these two engines could be *simpler?* In the middle there, with the Daytona 500 sweatshirt. Yes, ma'am.

DAYTONA 500: The engine used for the generator could be simpler.

FARNSWORTH: Why?

DAYTONA 500: Well, because the car engine has to operate over a wide variety of temperatures, so you need an exhaust manifold valve, and then you need a temperature-compensated choke in the injectors. Oh, and air pressure, too; the car has to be able to operate up high, so you have to have something to lean—

FARNSWORTH: Stop! Enough! That was wonderful. Exactly right. Now, could you tell me, which of the two engines would be more *reliable?*

DAYTONA 500: Well, the generator engine, of course.

FARNSWORTH: Why "of course"?

DAYTONA 500: Because it's simpler—there are fewer parts to go wrong.

FARNSWORTH: Marvelous! If you have a piece of machinery, you can make it more reliable if you can keep it simple. You can keep it simple if you narrow the range of conditions over which it has to operate. Now, is a cell a machine? Bet your boots it is. So if we want to keep the cellular machine as reliable as possible, what do we do? We try to keep the environment around the cell as constant as possible, no matter what is happening in the outside world. The maintenance of a constant internal environment for a cell is called **homeostasis**, and a lot of the control mechanisms of the body are geared to provide this homeostasis. Let's take a look at one of the variables, temperature, as an example of homeostatic mechanisms at work. First, what is the range of temperature extremes that a cell can tolerate? The high end first; what is the hottest a cell can tolerate? Any guesses?

STUDENT: Maybe the boiling point of water?

FARNSWORTH: Good guess; certainly if the water in the cell boils, the steam would rupture it. But is that really the highest point that a cell could tolerate? Think for a minute about boiling an egg. The white of the egg is protein, and when you boil an egg, something rather drastic happens to that protein—you can tell just by looking at it. But at what temperature does this "cooking" take place? If you were to boil an egg in Denver, where the air pressure is lower than at sea level because of its mile-high altitude, the water wouldn't boil at 212°F; it would boil at about 180°F. But you can still cook the egg, or in more technical terms, **denature**, or alter the structure of, the protein. Well, it turns out that most proteins will be fundamentally changed someplace in the range of 160–180°F. So is that the upper limit—the denaturation temperature of protein? Well, cells are made of things other than protein, so is there some other thing in the cell that would be adversely affected by temperature? How about the cell membrane? What is it made up of? Lipids, in the fat family. Butter. Butter is fat. What happens when you heat butter up to, say, 120°F, or so? It melts. Changes state. So what happens to the cell membrane at temperatures like that? It changes state, too, and so the cell is liable to rupture. So, someplace under about 120°F would seem to be the upper lethal limit for an ordinary cell in a multicellular organism. But wait a minute—has anybody here ever taken a sauna?

STUDENT: I have.

FARNSWORTH: And how hot was it in the sauna?

SAME STUDENT: Jeez, I don't know, maybe 175°.

FARNSWORTH: 175! So how come all your cells didn't blow up? Ah! The wonders of homeostasis. If you had been a rock, or dead, the temperature of your body would have soon come to 175°. But because you have homeostatic mechanisms, what happened? As soon as your skin temperature started to go up, your body detected that information, and almost instantly your sweat glands started pumping out a fluid that spread over your skin. Evaporation carried away not only the fluid, but a lot of the heat carried in the fluid. So the *internal* temperature your cells were exposed to remained constant.

Well, as we are going to see in a moment, it takes some very elaborate control mechanisms to maintain that homeostasis, but what do you do if you're an organism that for one reason or another doesn't have a lot of control mechanisms? Well, one option is to live in a place that has a relatively constant environment to start with. If we're talking temperature, the ocean's temperature doesn't fluctuate nearly so much as air temperature during the course of a year. So you have zillions of different kinds of invertebrate animals, like jellyfish or sponges, that don't have temperature control mechanisms; their cells can be kept happy, simple, and reliable because their *environment* is stable, just like the generator engine in its air-conditioned room. If you can move around, you can keep your internal environment relatively constant by behavioral means. Sauna too hot for you? Walk out of the sauna. Sun too hot for you? Crawl under a rock. The problem with the stable environment and behavioral mechanisms for producing homeostasis is that they limit your options. If you can operate in an unstable environment, and you don't have to crawl under a rock every time the sun comes out, there are a lot more places where you can live. So, you can buy flexibility by having internal mechanisms that can keep your cells stable. But you then have to control those mechanisms. Let's see how these mechanisms might work. (*He places a transparency on the projector.*)

Here is a cat on a hot tin roof. Heat can enter or leave its body in a variety of ways. The tin is hot, so heat enters its body by *conduction*, which is heat transfer by contact with a solid. If the cat were sitting on an ice cube, heat would leave—heat flows from hot to cold. If there's a cool breeze, the cat loses heat by *convection*, which is heat transfer by a fluid like air. This cat's a real heavy breather, and every time it breathes

out, it loses heat by the *evaporation* of water from its lungs. The sun is out, and the sun is hotter than the cat, so the cat gains heat by *radiation* from the sun. On the other hand, there is this giant ice sculpture of Christopher Columbus about 20 meters away, and it is colder than the cat, so the cat loses heat by radiation to ol' Chris. The cat just ate three mice and poor Mrs. Goldstein's parakeet, so the cat is generating heat by the digestion and *metabolism* of the unfortunate ex-mice and tweety bird. The total effect of all these energy transactions determines the internal temperature of the cat.

Now, in the interest of science, we are going to place this cat and a giant frog in a room in which we can control the temperature. I'm going to measure three things: room temperature, body temperature, and metabolic rate, which is a measure of how much energy each animal is releasing. I'll show you now a couple of graphs to indicate what happens. (*He places a transparency on the projector.*)

Look at the frog first. What happens to the frog's body temperature as the room temp goes down? It goes down, too. How about metabolism? It goes down, too. A rock doesn't metabolize, but the temperature of a rock in the chamber would vary about the same way as the frog.

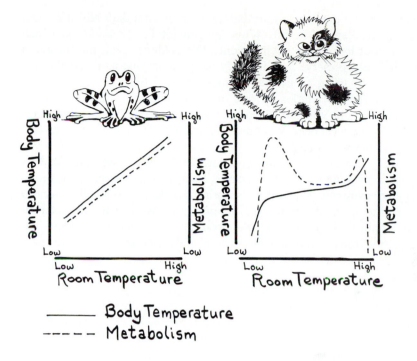

Body Temperature

Metabolism

Body Temperature

Metabolism

High — High   High — High

Low — Low   Low — Low

Low   High   Low   High
Room Temperature   Room Temperature

————— Body Temperature
— — — — Metabolism

Now look at the cat. A much more complex picture. Starting from the right, as the room temperature starts to go down, both temperature of the cat *and* its metabolism stay constant. A real puzzler. Then if I start to drop the temperature still more, what happens? The body temperature stays constant, but the metabolism starts to go *up*. There's a shocker—the frog's metabolism went down as the room temperature went down. If I continue to lower the room temperature, there's a point where the cat's temperature starts to go down, and then the metabolism starts to follow the temperature down. But the metabolism goes down to zero! What's happening here?

Well, for all practical purposes, the frog is really acting thermally like a rock. Its body temperature starts to drop, and as this happens its metabolism drops, too. Why? Well, metabolism is a series of chemical processes, and the rate at which chemical processes go on is a function of their temperature. So if the temperature goes down, the metabolism slows. What is the practical significance of this for the frog? As the environmental temperature drops, the frog gets more and more sluggish. He can't jump after flies, and he can't avoid snakes. So what does he do in cold weather? Something very sensible; he buries himself in mud for the winter, and doesn't come out 'til it warms up. He can get away with this because his metabolic rate is so low that he doesn't need to input any food energy for months.

Now consider the cat. If we start over at the right of the graph and go left, what happens? Well, as the air gets colder, the cat is going to lose more and more heat to the air. If the cat were like the frog, its body temperature would drop. But it isn't—look at the graph. How can the body temperature stay constant in the face of lowering environmental temperature? Well, have you ever seen a long-haired cat on a really cold day? Is it the same *size* as the same cat on a hot day? No! It's much bigger because it has elevated its fur. Why? To provide better *insulation,* which slows the rate of heat loss to the environment. The colder it gets, the thicker it makes its fur. So it can preserve both its body temperature *and* its metabolic rate by physical means—by using a variable thickness of insulation.

But now, I crank the refrigeration up, and the cat has elevated its fur as much as it can. Now what does it do? What do *you* do when you get real cold? You shiver, that's what—and that's what the cat does. What does shivering do for you? Well, a shiver is a powerful muscle contraction that doesn't result in movement of a limb. When you contract a muscle, you burn food. When you burn food, you release heat. Aha! By shivering, the cat is cranking *up* its metabolic rate, releasing heat to replace the heat it is now losing to the environment at an ever-increasing rate. So the cat is still preserving a constant internal temperature, at the cost of a greater energy expenditure than the frog, which for all practical purposes has shut down for the winter at this point. The cat, on the other hand, can prowl around, avoid dogs, and eat canaries—but note this— it *has* to eat canaries, because it is burning up energy at a fierce rate, whereas the frog is just snoozing.

So now, the cat is shivering like mad and has its fur fully elevated, and I decide to take the temperature down another couple of degrees. What does the cat do now? Nothing—it has run out of options. It is losing heat faster than it can generate it, and so its body temperature starts to drop. When that happens, its metabolic rate drops off, which means it releases *less* heat than it did a couple of degrees ago. This lowers its body temperature even more. Body temp and metabolism chase each other down until the body temperature is the same as the environmental temperature, and metabolism goes to zero because the cat is dead. Now, I hope you realize that I would never do this experiment myself, because I sort of like cats, but it is very important to know how this mechanism works. A warm-blooded animal like a cat or a human can maintain full activity over an amazingly wide variety of environmental temperatures, but once the temperature drops below a certain level, death can come very rapidly, and that information would be very handy in a survival situation.

My point in giving you this demonstration is to show you that an animal which has adapted to operate in a wide variety of environmental conditions needs to provide homeostatic conditions for its cells and must have a fairly elaborate control system to maintain that homeostasis. But think of the complexity of all this! Temperature. Oxygen level in the blood. Sugar balance in the blood. The number of things you have to control is almost endless. Staggering. If you had to think consciously about all these things, it would be impossible. So what you have to have is some kind of control mechanism that can take care of all this automatically. There is such a mechanism, and it is called *feedback*. It is probably the single most important concept I'll teach you all semester. If you understand feedback, you will understand not only how biology works but also why nations go to war, why the dollar is worth what it is, and why the landlord can't raise your rent indefinitely.

Normally, I like to talk about concrete things first, and then go to abstracts, but this situation is just a little bit different, and we will need some vocabulary to discuss the concretes.

First of all, an ***input-output device*** is anything which produces a product and whose rate of production of that product is influenced by another product. Makes no sense at all, does it? Short-order cooks are input-output devices. They produce pancakes, a product, and the number of pancakes they make depends on the order slips, another product, that the waiter or waitress gives them. The product can be either a tangible, physical thing like a pancake, or information, like the number of pancakes listed on the slip. The input could be physical or informational; doesn't matter. The same is true for the output. A phone is an input-output device. You give it a number, and it gives you the voice of the person you want to talk to.

If we have *two* input-output devices, we can connect them in a particular way so as to give us something called a ***feedback loop***. A feedback loop is two input-output devices connected so that the output of one influences the output of the other, and vice versa. Another way of putting it is that the output of one is the input of the other. Let me show you. (*He draws two feedback loops on a transparency.*)

In the top loop, *A* is a general input-output device. The arrow going out is the output; the arrow coming in is the input. Over to the right we have another input-output device, *B*. Device B has an output, and you can see that B's output is A's input, and vice versa. We have now created a *loop*, in which both devices influence each other.

The first specific kind of loop is called a ***positive feedback loop***. *Positive* doesn't mean "good" here. What it means is that the output of the first input-output device *stimulates* the second, causing it to produce more of its output. The effect of the

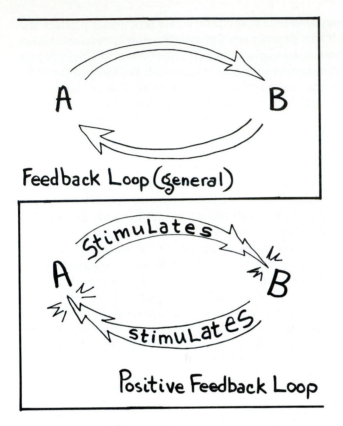

Feedback Loop (general)

Positive Feedback Loop

output of the second device is to *stimulate* production of the output of the first. In other words, they stimulate each other. Example time now.

College guy and gal sitting on opposite ends of a couch. They both like each other, but they're both kind of shy and waiting for a first move from the other. Neither of them has an output at this point—this is the initial condition.

Something happens—he gets an itch, and in trying to scratch it he shifts position on the couch and moves a half-inch closer to her. Now he has produced an output—which becomes her input. She sees the movement, thinks "Maybe he *does* like me," and moves an inch closer to him. Now *she* has produced an output, which becomes his input. He sees the movement, thinks "Maybe she *does* like me," and moves 2 inches closer. She sees this, thinks "He *does*! He does like me!" and slides over the remaining distance. He now thinks "Likes me? Hell, she *loves* me!" and they lock lips. *She* now thinks—well, maybe this would be a good time to discreetly draw the curtain on our young friends.

The important point here is that once started in motion, a positive feedback loop tends to *accelerate* the production of the two devices until both of them are operating at *maximum output*. A biological example might be a fever. You start out with a stable temperature—in this case *temperature* is an input-output device because it influences other things, and other things influence it. Something happens, you get an infection, for example, and your temperature goes up a little. What happens to your metabolism when temperature goes up? Metabolism goes up, too—the reactions of metabolism speed up with increased temperature. As your metabolism increases, what happens to your heat output? Well, remember the cat, the higher the metabolism, the more heat is put out. What is the effect of this heat on your temperature? Clearly, it goes up. And what does that do to metabolism? Drives it up. Each turn of the "wheel," or circuit around the loop, causes each of the "devices"— temperature and metabolism—to increase its output.

One of the properties of positive feedback loops is that they accelerate. Another property is that once started, the devices go to maximum output—unless something breaks the loop. You take a cold bath to knock the temperature down, for example.

I should point out a potential trap here. *Positive* suggests going in an upward direction. Not so, necessarily, in a positive feedback loop. For example, you're driving a car at 80, and you step on the brake. The car starts to slow, and momentum drives you forward onto the brake pedal, pushing it harder. The car decelerates even more (*deceleration* is negative acceleration), and the loop keeps going until you are braking as hard as you can, and the car is stopping as hard as it can. *Positive* just refers to mutual *stimulation*.

Positive feedback loops aren't very important in biology because they don't really *regulate* things, but the other kind of loop, the **negative feedback loop**, is extraordinarily important. In a negative loop, you have the two devices, A and B, just as before. The output of A stimulates the output of B, just as before. But the output of B *reduces* the output of A, and there is the critical difference. Suppose we have our loop going, and A is producing its product at a constant rate. Now some outside force causes it to increase its output. What happens? B gets stimulated to put out more of its output. What happens when the output of B gets received by A? A is *inhibited*. The outside force kicked A's output up higher than it was before, but now B pushes A's output down to normal. With a normal output from A, B is no longer stimulated, and its output goes back to normal too. So the system is *stable* again.

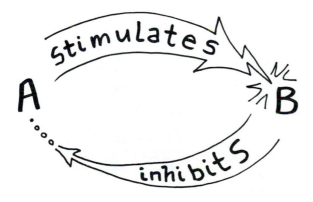

Negative Feedback Loop

Let's look at a practical example. What temperature is it in this auditorium? Anybody, a guess?

STUDENT: About 45°. (*Laughter*)

FARNSWORTH: All right, all right. The school is cheap, but not *that* cheap. Probably 65°F. And will it stay at that temperature all day? Pretty close. But how about the outside temperature? Is it constant? No. So why doesn't the temperature of the room follow the outside temperature? Is there a little man in the basement with a thermometer who goes outside every 10 minutes and cranks the furnace up or down? No. What do you have instead of the little man? A thermostat.

There's the thermostat on the wall. It is an input-output device. Its input is the room temperature—information, in other words. Its output is an electrical signal that stimulates the furnace to turn on. The furnace is the other input-output device. Its input is the signal from the thermostat, and its output is heat that raises the temperature in the room.

Okay, let's start the system out. The outside temperature is 65°, so we set the thermostat at 65°. It's not sending anything to the furnace, and the furnace is not sending anything to the thermostat. The outside temperature now drops, and the temperature in the room starts to drop. What does the thermostat now do? It sends a signal to the furnace—*stimulating* it to turn on. So the furnace starts cranking out heat. What is the effect on the thermostat? As soon as the temperature hits 65°, the thermostat is *inhibited* in its output and no longer sends a signal to the furnace. The furnace, no longer being stimulated by the thermostat, shuts down, so the temperature in the room doesn't shoot above 65°. So now both the thermostat and the furnace shut down when the temperature is at the *set point*—the level of the variable, in this case temperature, that you want to stabilize. So within the limits set by the size of the furnace and the outside temperature (it might get so cold that the furnace couldn't keep up with the heat loss)

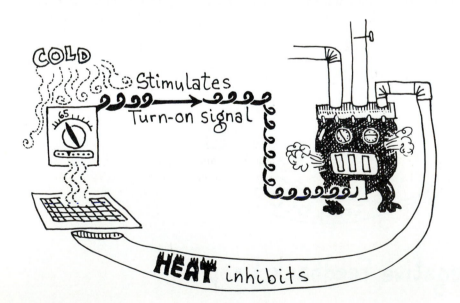

the negative feedback loop represented by the thermostat and furnace will tend to keep the temperature constant around the set point.

Another quick example. Landlords. The average rent in your area is $400 a month, let's say. Your landlord has 10 tenants, so his income is $4000 a month. He gets greedy and raises the rent to $800 a month. Is his income now $8000 a month? Only for the first month because six of his tenants move out and he can't get new ones. So his greed, which was stimulated by the extra income, has now been inhibited by fear of losing all his tenants. So he lowers the rent back down to $400 and fills up again. The set point of his rents is determined by the availability of alternatives to his tenants. Society is full of negative feedback loops. Labor and management. Labor can't raise its wage demands too much, or the company will go out of business. Management can't squeeze the workers totally dry, because the company will lose all its workers. The set point for wages is determined by the prevailing average wage. Now, if the average wage goes up, because of inflation or whatever, the set point will be changed and wages in a given company can rise, just as you can change the setting on the thermostat to a higher level, causing the temperature to rise and stabilize around the new set point.

Biology is chock-full of examples of negative feedback loops. The numbers of foxes, or any other predator in an area, can't go up indefinitely because they'll eat up all the rabbits. The foxes then will starve down to the number of foxes that can be supported by the rabbits in the area.

You drink a glass of water. It is important to keep the volume of your blood constant, for homeostasis. So what happens? You absorb the water into your bloodstream. Your blood volume is now greater because blood is mostly water. Your blood also becomes more dilute, in terms of its salt concentration. There are sensors in the bloodstream that sense this and send the message to the pituitary gland and the brain. They in turn send a message to the kidney, telling it to excrete more water. You get an urge to urinate, and pee out the excess water. Your blood volume and concentration are back to normal, the sensors pick up that information, tell the brain, and the brain inhibits the kidney from any more urination. There is thus a negative feedback loop between the pituitary and the kidney to keep blood volume and concentration constant. In a biological system that needs to have something kept constant, always be on the lookout for a negative feedback loop. The clue is two input-output devices: one stimulates; the other inhibits.

Think about this business with feedback loops, and you will be able to think of dozens of examples of things in everyday life, in every arena of life, that are regulated by negative feedback loops.

Well, I don't know about you, but this has been a full day for me. Bruce and Dick and I are going to go back to the lab now and collapse. (*He swiftly leaves the podium so students won't be inclined to go up to the stage and find the trick cage on the podium. He'll return after the class is gone to retrieve it.*)

Day 16

Nerve Impulses and Countercurrent Systems

Professor Farnsworth enters wearing khakis and a blue button-down shirt. He carries a large gym bag.

FARNSWORTH: Good morning. This is going to be a sort of miscellaneous lecture. You're reading your physiology chapters in the text about now, and I've found that there are certain ideas that always seem to cause problems, and can benefit from body language in the explanation. These concepts aren't necessarily the most important ones, but they are either a little complex or a little nonintuitive. Don't look for any thread tying things together in today's session—they're separate ideas.

We'll start with the nervous system, specifically, something called the *nerve impulse*. A *nerve* is a structure that carries information. If it carries information from somewhere to a central processing facility, like the brain, it is called a *sensory nerve*. If it carries information from the central location to another structure where something happens, it is called a *motor nerve*. A nerve is made of a series of long, skinny cells called *neurons*. I'd better draw you one here. The main part of the cell is called the *cell body* and contains the nucleus. The whole cell is kind of *amoeboid;* that is, it doesn't have a well-defined shape. There is a series of short, branching projections on one side of the cell, which are called *dendrites*. On the other side is a long, fingerlike projection called an *axon*. There's a lot of variation on this pattern, but this is close enough for a general impression.

The end of this axon is kind of interesting, so I'll show you a closeup picture here. (*He places a transparency on the overhead projector.*) The axon splits up into short branches, and at the end of each little branch is a swelling called the *synaptic knob*. Inside the synaptic knob are a bunch of little containers called *synaptic vesicles*. These synaptic knobs come very close to, but don't actually physically touch, another neuron, or in some cases a muscle or a gland. If the synaptic knob is close to another nerve, the junction between the two neurons is called a *synapse* and the space between the two is called the *synaptic cleft*.

Okay, suppose we want to send a message along a nerve, what do we do? Let's say I accidentally stuck my finger in a flame. What would happen? Well, clearly,

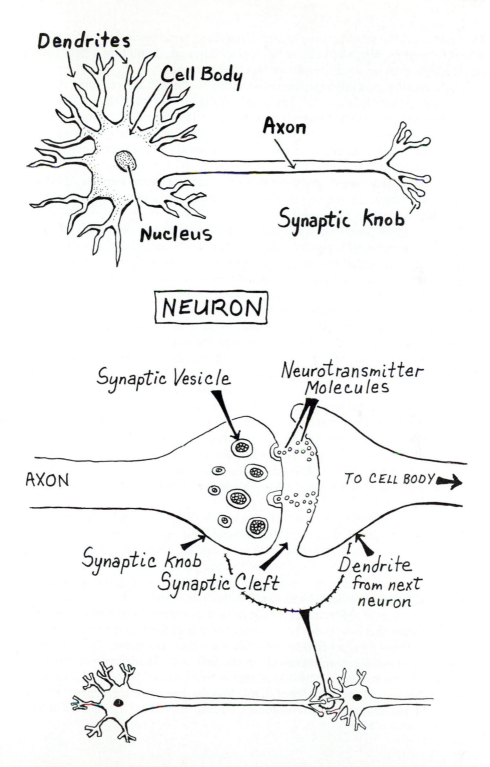

NEURON

I'd jump away, but would I do it the *instant* the finger went in the flame? No. There would be a delay. Why would there be a delay? After all, if you turn a light switch on, the light goes on instantly, for all practical purposes.

The reason there is a difference between turning on the light and sending a message along a nerve is very fundamental. In turning on a light, you are dealing with an *electrical* phenomenon. Electricity travels along a wire at 186,000 miles an hour, so for all intents and purposes, the instant you hit the switch and electricity starts flowing down the wire, the light will go on. Transmission of a signal along a nerve is *electrochemical* rather than electrical, and goes much slower—about 300 feet a second.

The message that travels along the neuron is called a **nerve impulse**. This impulse is not a physical thing that carries information, like a letter or a telegram. It is more like a wave on the ocean.

A nerve impulse starts when the end of a sensory nerve is stimulated by something. That "something" might be light, pressure, heat, cold, chemicals, or a whole variety of other potential stimulants. Once the nerve has been stimulated, a nerve impulse passes down the length of a neuron in a way that I'll describe in just a moment. When the impulse reaches the synaptic knob, the synaptic vesicles release substances called **neurotransmitters** which diffuse across the synaptic cleft. The effect of these neurotransmitters is to stimulate the next neuron in line, which then passes a new nerve impulse along. Eventually the impulse reaches the brain, or other central processor, and the meaning of the impulse is interpreted. It's something like the brain's saying, "Oh, here comes a nerve impulse from a temperature sensor in the guy's finger. Wow! It's hot! Tell the finger to pull away." A motor nerve is then stimulated by the central processor, and a nerve impulse passes from neuron to neuron until it reaches a muscle cell. The effect of the neurotransmitter at the end of the last motor neuron is to stimulate the muscle to contract.

I will do a demonstration of the nerve impulse now, showing its electrochemical, rather than electrical, nature. This demonstration is undoubtedly overkill—it's not that difficult an idea—but you ought to know me by now, I can't resist a flashy demonstration.

Allow me to call your attention to the back of the stage. You will notice that there are two 25-foot rolls of aluminum foil stretched out on the floor, with the ends joined at one end to form a shallow *V*. As you can see, there are five short lengths of aluminum foil spread out from one of the top parts of the V. These represent fingers at the end of an arm. At the end of the other leg of the V there is a 3-foot-square flat plate of steel. I will stand on that plate, and my body will represent a finger muscle. Starting from the tip of one of the "fingers" you will notice a trail of black powder that runs about half-way up one side of the V and ends in a mound of powder. About 3 inches farther along the V another trail starts. There are three of these trail-mound-gap arrangements in the first arm. At the intersection of the legs of the V you will see a network of intersecting trails of powder, then another mound at the start of the other leg of the V. Moving down that leg, there is another sequence of trail-mound-gap arrangements, leading to a rather large mound of powder on the steel plate.

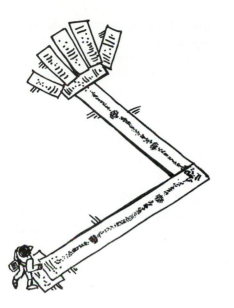

The powder is ordinary gunpowder, which in the mounds is mixed with flakes of magnesium metal. When all is ready, I will ask one of you to touch a match to the powder trail at the tip of one of the fingers. That will represent a heat stimulus to the end of a sensory nerve ending. The heat of the match will start a chemical reaction in the gunpowder, which will result in the output of heat. That heat will initiate chemical reactions in the powder a little farther along in the "neuron," and you will see this simulated nerve impulse move, not instantaneously as would be the case with electricity along a wire, but at a brisk, reasonable speed.

When the "nerve impulse" reaches the mound of powder which represents the synaptic knob, the magnesium flakes will represent the synaptic vesicles. When they ignite, the burning flakes will represent the neurotransmitters, which will jump the synaptic gap and then ignite the powder trail in the next neuron. At the intersection of the V, the network of trails represents the pathway of neurons in the brain. Ultimately, the flame will find its way out of the brain and down the other leg of the V, which represents a motor nerve. Eventually, it will reach the plate, which represents the neuromuscular junction, and since I represent the muscle, you should see a fairly rapid response. You will note a large mound of powder on the plate upon which I will stand. Now, I know some of you would like to see your ol' professor go up in flames, but I am going to take this small precaution before I begin.

He pulls a Nomex fireproof racing suit out of the gym bag and dons it. While he is doing this, several of his graduate student assistants appear with fire extinguishers. When he has the suit zipped up, he steps over to the plate and signals one of the grad assistants to give a box of matches to a student sitting in the front row. The student walks over to the finger of aluminum foil, strikes a match, and drops it on the powder trail, which immediately ignites. The flame moves at about a walking pace down the trail until it hits the first mound. When the mound ignites there is a shower of sparks, some of which jump the "synaptic gap" and land on the beginning of the next powder trail. Farnsworth stands without moving as the flame moves toward the "brain." At the brain, the flame divides and races up and down the intersecting trails, until one flame front reaches the "motor nerve" trail. It then passes down the motor nerve until it reaches the steel plate upon which Farnsworth is standing. There is a tremendous flash, and a huge cloud of smoke instantly envelops him. When the smoke clears, Farnsworth is seen sitting on his rear end a couple of feet away from the plate, in a posture which makes it look as though he was actually blown off the plate. He slowly stands up, pulls the hood off the suit, and begins to unzip it as he speaks.

FARNSWORTH: Don't try this experiment at home! *That* will wake you up in the morning! Well, this somewhat gratuitous demonstration should have suggested the general principle of the nerve impulse. Now I will show you how the details work. (*He goes to the overhead projector and starts to draw.*)

Remember ions? Positive ions like potassium, sodium and hydrogen? Negative ions like chloride? These are *charged particles* because they carry electrical charges. Now, if you have a positive and a negative charge located close to each other, you have what's called a **potential difference** between them, or a difference in electrical potential, which is measured in volts.

(*He places a drawing on the projector.*) Okay, I have an imaginary pot of water here, and I dump a bunch of sodium ions in the water, so now I have all these

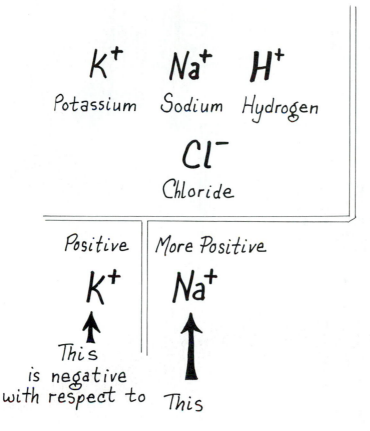

positively charged ions floating around. Now I get a water balloon, but instead of filling it with tap water, I fill this ballon with water in which I have dissolved some potassium ions. I toss the balloon, which represents a neuron, into the pot of water, which represents the fluid surrounding the neuron. Most of your body's cells are surrounded by this fluid, which is called the *tissue fluid*. (You may have seen this fluid if you've had a shallow paper cut. It might not have bled, but you could see a sort of clear fluid oozing out of the cut.) The skin of the balloon represents the membrane of the neuron.

We now have the balloon in the water. There's sodium outside the balloon, potassium inside, and both are positively charged. But this is a slightly defective balloon. It has a whole bunch of tiny holes, just big enough to let potassium through, but not sodium. So what happens is that the potassium starts to diffuse out of the cell. As it does, since opposite charges attract each other, any stray negatively charged particles will tend to be drawn toward the inside of the membrane. If we now used a voltmeter in this situation, with one probe on inside of the membrane, one probe on the outside, what do you think we'd find?

STUDENT: The inside of the membrane would be negative with respect to the outside.

FARNSWORTH: Exactly, The inside of the membrane is negative to the outside because you have negatively charged particles clinging to the inside of the membrane, and because you have a slight increase in the number of positive charges on the outside of the membrane, as some of the potassium leaked out.

Back to our voltmeter. If we actually measured the potential difference between the two sides of the skin, we'd see that the inside was a little less than a tenth of a volt negative with respect to the outside. More precisely, we would say it was -70 millivolts, or thousandths of a volt, negative. This negative potential in a nerve is called the *resting potential*. So, when the nerve is in a nonstimulated state, it has the resting potential.

Okay, another memory refresher. Remember osmosis? You have a membrane that can let some ions through? What would happen if the balloon skin were permeable to sodium?

STUDENT: Since the concentration of sodium outside is greater than inside, you'd have a net flow of sodium to the inside until there was as much sodium on the inside as on the outside.

FARNSWORTH: Right. At the moment, we have what's called a *concentration gradient* of the ions. More sodium outside than in, more potassium in than out. But what if the membrane is full of holes? Would you expect to maintain a concentration gradient and a resting potential then?

STUDENT: No, because the two ions would diffuse back and forth until they both were in equilibrium on either side of the membrane.

FARNSWORTH: Aha! That is exactly what you would *expect*, but that isn't what you find in a neuron. The membrane *is* permeable, but there is still a concentration gradient of both sodium and potassium. How is this magic performed? The idea is simple; the execution of the idea is complex. You have to *pump* the sodiums outside

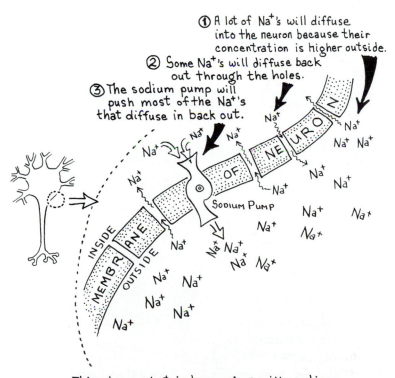

① A lot of Na⁺'s will diffuse into the neuron because their concentration is higher outside.

② Some Na⁺'s will diffuse back out through the holes.

③ The sodium pump will push most of the Na⁺'s that diffuse in back out.

This shows what is happening with sodium. The same thing is happening with potassium, except in the *reverse direction* — the potassium is mostly inside and diffuses out.

and the potassiums inside. You've got a big concentration difference from outside to inside. Now you open up the holes in the membrane to make it permeable. A bunch of sodiums start flowing in, and potassiums out. But here's the clever part. *Inside* the membrane there is a chemical pump, called the **sodium-potassium pump**, which grabs sodium ions that end up on the inside of the membrane, after they have flowed into the neuron by diffusion, and pumps them back outside. The pump does the same thing for potassium in the opposite direction. So what happens is that as fast as the sodiums flow in, and the potassiums flow out, the pumps grab them and kick them back where they came from. This way a concentration gradient is maintained, even though the membrane is permeable.

There's one other thing you need to realize about this pump. Like any pump, it requires energy to operate, in this case ATP. When you move ions against a concentration gradient, with an expenditure of energy, you have what is called *active transport*. So the concentration gradient and the resting potential are maintained by active transport.

Now we're going to make our balloon more like a real neuron. (*He places a new transparency on the projector.*) First, we stretch it out so that it is as long as an axon.

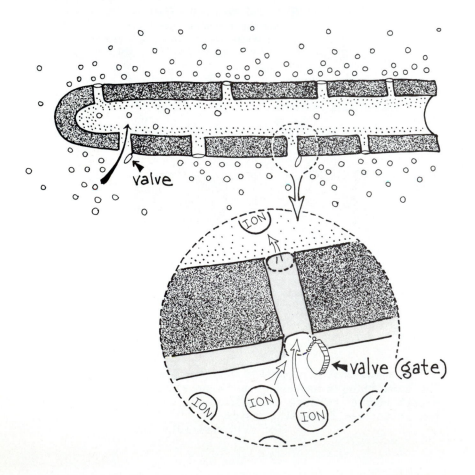

Next, we're not going to have simple holes punched through the membrane; we're going to have valves, or gates, that we can open or close to let ions pass through. Now, we're going to stimulate it at one end with something and see what happens at the point of stimulation.

What happens when you stimulate a nerve is that at the point of the stimulation—and the stimulation can be almost anything (heat, light, pressure, chemicals, etc.)—at that point, the gates open up and the positive ions can move through the membrane. Sodium starts to diffuse in, and potassium starts to diffuse out. What do you suppose happens to the resting potential? It starts to change—I'll show it here on a little graph. The inside of the cell was negative with respect to the outside before the stimulus, but now that you have strongly positive sodiums moving in, the inside becomes positive with respect to the outside. At that point, the neuron is said to be *depolarized*. So now you have these open gates, and ions flowing. You have now generated what is called an *action potential*, a voltage originally created by a stimulus. A very interesting thing happens next. We started this action potential by applying a stimulus right at the end of the neuron, and the gates in the immediate area of the stimulus opened. The effect of having these gates open is to cause the gates right next to the open gates to open up, too. So now, we've moved a little bit down the neuron, and we have ions flowing here, too. Well, once these gates have

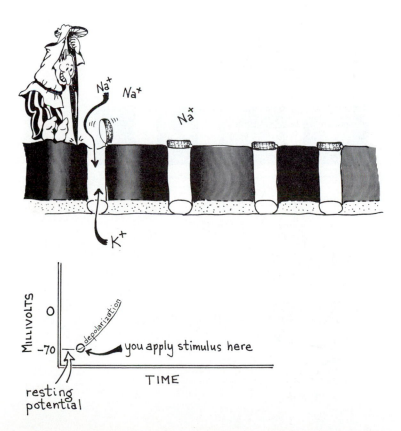

opened, the gates next to *them* open up and, in turn, the gates next to them open up. So what you have is a *wave* of gate opening that moves down the neuron. This wave of gate opening, and the other events associated with gate opening that we'll talk about in a second, is the nerve impulse. The nerve impulse is an action potential that moves down the neuron.

Let's go back to the point of the original stimulus. The neuron is now depolarized at the site of stimulation, gates are beginning to open next to the original site, and we now have more sodiums inside and potassiums outside. Two things now happen. The gates at the original site snap shut, and the sodium-potassium pumps in the vicinity of the original stimulus kick in and start pumping ions back against the concentration gradient. What is the effect of this?

First, the voltage begins to swing back toward the state in which the inside of the membrane is negative. This is called **repolarization**. Repolarization continues until the inside is back to −70 millivolts, but the pumps don't kick out right away. They overshoot a little, and the voltage then settles back down to −70. This overshoot is called **hyperpolarization**.

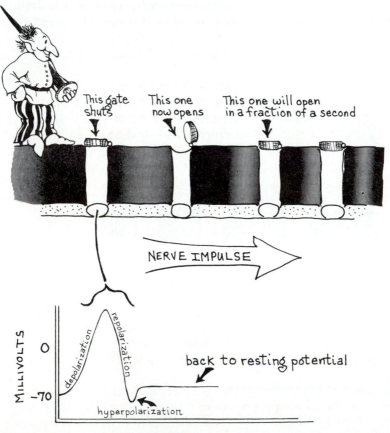

Now look at what is happening to the gates next to the repolarizing gates at the site of the original stimulus. The effect of having the pumps kicking in and the gates shutting down next door is to cause the same thing to happen to the neighboring gates and pumps. So what happens is you get a wave of depolarization, repolarization, and hyperpolarization that moves down the axon. As soon as this wave hits the synaptic knob, the change in concentration of ions at the knob produced by the opening of ion gates in the knob causes the synaptic vesicles to pop, releasing their neurotransmitters into the synaptic cleft. These neurotransmitters act as stimulants for the next nerve in line, or muscle, or gland, or whatever.

Okay, that's the nerve impulse. The next idea I want to talk about is **countercurrent flow**. I can explain what it is best by giving you a problem. Have you ever seen a duck standing on a frozen pond? Yeah, all right, some of you have. Now, this duck has a problem. He has to have blood going down to those stupid-looking feet, but he's a warm-blooded animal, right? He's going to lose a tremendous amount of heat to the ice because the blood passing from his body into his legs is hot. The bigger the temperature difference is between his feet and the ice, the more heat he loses. So if he has hot feet, he's going to lose all his body heat to the ice. Nasty situation. What he wants to do then is have some kind of mechanism that will pull the heat out of the blood and send it back to the body, before it ever comes close to the ice. He wants *cool* feet, man. He does this by arranging the artery going down the leg right smack next to the vein bringing blood back from the feet. (*He puts a transparency on the projector.*)

The drawing shows two feet, one arranged with countercurrent flow; that is, the fluid flowing one way in a pipe is right next to a pipe carrying fluid in the other direction. That's why it's called "countercurrent." In the other foot there is no contact between the two pipes—no countercurrent flow. In the countercurrent foot, the blood coming out of the body of the duck into the legs is at about 40°C. Heat can flow out of the artery and through the skin of the leg to the outside, but more of it is going to flow into the blood in the vein going back to the body. A little farther on in the artery, we see that the temperature has cooled off quite a bit, but most of that heat has been returned to the body. By the time you get to the feet, the blood isn't that much warmer than the ice, so not that much heat is lost to it.

Over in the leg on the right, the blood starts out at the same temperature, but since there is no direct contact with the blood going back to the body, the blood stays hot in the artery until it gets to those webbed feet. Now, you've got these hot feet, and there is a tremendous loss of heat to the ice.

Okay, you find this countercurrent exchange arrangement in places where you want to keep heat inside the body, rather than allow it to leak out through the limbs. Also, some fish, like tuna, have a countercurrent arrangement to keep the center of the body warmer than the surface, again to prevent loss of heat to the environment.

Another place you can find a countercurrent mechanism is in the gills of fish. Blood from the body that needs oxygen moves through the small blood vessels of the gills in a direction opposite to the flow of the highly oxygenated water that crosses the fish's gills. By the time the blood has reached the far end of these small

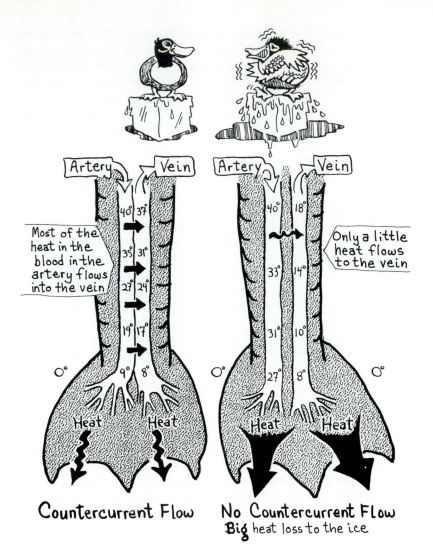

Most of the heat in the blood in the artery flows into the vein

Only a little heat flows to the vein

Countercurrent Flow

No Countercurrent Flow
Big heat loss to the ice

vessels and is ready to enter the blood vessels returning to the body, its oxygen concentration is almost as high as that of the water entering the gills. In this case oxygen is exchanged instead of heat, and the transfer is between water and blood rather than between two blood vessels, but the principle is the same as with the duck. Think about what would happen if the water and blood flowed in the *same* direction.

Okay, those are the two biggies, nerve impulses and countercurrent mechanisms, that always seem to cause the most problems. I think we can quit while we're ahead.

You have been working very hard so far, and I'm very pleased with the class, so I have a special treat for you. I will not be here for the next lecture (*cheers from the class*). Ha! That's gratitude for you! Anyway, an old professional colleague of mine, Dr. Viktor Alucard, of the Bucharest Institute of Hematological Sciences—look up *hematology*—will give the lecture, and his topic will be "The Biological Basis of Myth and Legend." In answer to your unstated questions, yes, it will be on the exam, and yes, the notes will be complex, so it is best to take them yourselves. I'll see you lecture after next. (*He exits.*)

Day 17

The Biological Basis of Myth and Legend

It is a Friday toward the end of October. When the students have seated themselves, the auditorium lights fade to black and the ominous strains of the "Todentanz," or "Dance of the Dead," rise on the huge PA speakers. A spotlight picks out the left entrance to the auditorium. The doors pull open by unseen hands, and six hooded figures, perhaps monks, enter with measured steps, carrying a black coffin. They walk to the center of the stage, gently set the coffin on a black catafalque in the center of the stage, and disperse to the front seats of the class. The spotlight shines directly at the head of the coffin. The music fades to silence, then the howl of a wolf is heard. The coffin lid cracks open a hair, and four ghastly pale fingers grip the edge of the casket. Slowly, slowly, the coffin lid is pushed open. A head appears in profile, its eyes closed. A man rises from the waist, his torso now visible above the coffin. The side of the coffin facing the class silently drops away, permitting the figure to swing his legs over the edge. The figure stands and faces the class. His cheeks are hollow, and his skin is almost white, but his lips are startlingly red. His black hair is slicked back with brilliantine. He is wearing white tie and tails and a gorgeous black silk cape, lined with crimson satin and trimmed with silver buckles. It is Farnsworth, of course, but the disguise is complete because the figure is clean-shaven. Almost none of the students recognize him, so they don't know how to respond. Is this going to be funny? Is this serious? What is happening?

The figure opens his eyelids to reveal light-blue eyes rimmed with red. Slowly, as if awakening from a long, long sleep, the figure walks to the podium and begins to speak. His accent is a very heavy central European one, and difficult to understand, so it will be indicated only in the first paragraph below, to give the flavor of his speech. Subsequent paragraphs will be unaccented, translated into normal English for easier reading.

ALUCARD: Gut mornink. Eet iss gut to see you. Eet iss gut to see ANYONE after a night in ze tomb. Zo damp in zere, and cramped, und all ze little animals, crawling across your face. Terrible. However, you are not here to leesten to my problems. Your mentor, ze—*late*—Professor Farnsworth (*nervous laughter from class*) has asked me to zpeek about a topic upon which I haf a zertain—expertise. Und zat iss ze biological basis of myth und legend. Vat iss ze TRUT about ze

244

famous stories uf ze Loch Ness—monster—ze zerpents, verevolfes, zombies, und my favorite, heh, ze VAMPIRE! (*He swirls the cape around in an exaggerated gesture of emphasis.*)

Before I begin, allow me to introduce myself. I am Cou—, er, Doctor Viktor Alucard, of the Bucharest Institute of Hematological Science, and I suppose I do owe you an explanation for my somewhat curious appearance. You see, I have a very rare genetic disease, which causes me to be extraordinarily sensitive to light. The slightest touch of sunlight would be instantly fatal to me, so to prevent that from happening, I have been forced to always travel in this box during daylight. I have a little royal—blood—in my veins, and these men who carry the box are descendants of families who have been in my family's service for hundreds of years. And as to my attire (*swirls the cape*), in Europe a lecture is a formal occasion, I am a formal man, and thus the formal dress is appropriate. So, you see, there is nothing—*supernatural*—about me, is there? (*He is totally unconvincing in his last statement.*)

With that out of the way, let us get down to business. What we are going to talk about today is the intersection of biology, anthropology, psychology, and the supernatural. Science cannot say whether there are or are not supernatural things—the tools of science are simply not equipped to answer such questions. But the *belief* in things supernatural is very strong among people. It is my thesis that many of the stories that are considered supernatural today had their origins in observations of real natural phenomena at some time in the past, and by a process very similar to the "rumor game," a popular party activity, these real-world observations got transformed into myths and legends.

For those of you unfamiliar with the rumor game, a group of people sits in a circle. One person whispers a made-up story to the person to the right. That person passes the story in turn to the next person to the right until the story passes around the circle. Even though people are instructed to repeat the story exactly as they heard it, inevitably the story will be changed, and the most common type of change is exaggeration. At the beginning, the statement might have been "He caught a big fish." By the end of the story's passage, it will have been changed into "He caught a whale!"

To give an example of this, in the Old Testament, in Genesis 6:4, there is a statement to the effect that in the "old days" there were giants on the face of the earth. In the story of David and Goliath, Goliath was a giant, whose height when translated into present-day measurements would be over 20 feet. Was Goliath 20 feet tall? Possibly, but it is far more probable that Goliath was an extremely tall ordinary man, perhaps as tall as the 8 and some odd feet of the tallest contemporary man. Suppose you were a superstitious middle eastern villager living in biblical times who visited a neighboring village and saw an 8-foot-tall man. Would that make a strong impression on you? Absolutely. Now you return to your home village to tell the story. "A giant! A giant lives in Yattah!" A skeptic then asks, "A giant! You must be joking. How tall was he?" Defensively, you reply, "Huge! He must be at least 10 feet tall." Well, 10 feet tall is not so absolutely farfetched that it is impossible to believe, so the second man accepts your story; then when he in turn visits another village, he slightly embellishes it. At each turn of the tale, the giant gets bigger.

Remember, most of these legends developed at a time when there was no photography, no TV, no way of making an indisputable record of what you saw. Belief was entirely dependent on the credibility of a witness. For example—

As he talks, he casually reaches into his jacket and pulls out a gold cigarette case. He extracts a cigarette and places it between his lips. His right hand is clearly visible to the class at all times. He snaps the fingers of his right hand, and a small flame shoots out of his index finger. He lights the cigarette, then blows on the tip of his finger. The flame goes out.

ALUCARD: Some of you saw me light the cigarette with my fingertip. In today's scientific times, you might say, "Well, it is just a trick; he has some kind of chemical on his fingertip." But suppose you don't know any science. How would you explain what I just did, *especially if you did not think I was a magician?* You would be almost forced into a supernatural explanation. In the present case, neither explanation is correct. I do not have some high tech chemical on my finger, nor am I an agent of Satan. The explanation was right in front of you, but you were not prepared to see it.

With this in mind, let us look at some myths, and see if we can figure out how they might have started. Let us examine first the Loch Ness monster.

Loch Ness is a lake in Scotland, 27 miles long, a mile wide, and very importantly, about 700 feet deep. There are abundant underwater ledges and caves, where

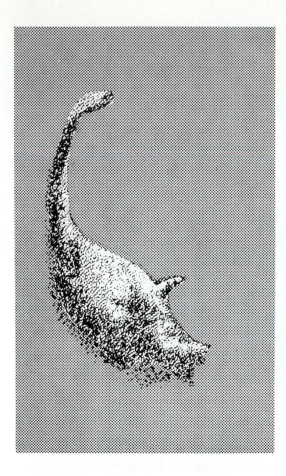

something large could hide. The water is extremely cloudy with peat—so cloudy
that you can't see your hand if you stick it in the water. This brings out an important thing about the origins of myths. They originate in situations which make
it difficult to disprove them. Why don't we settle the Loch Ness monster question
once and for all by using underwater cameras? Because the water is so cloudy that
cameras are useless. The ancients didn't have cameras, but they did know the lake
was very, very deep, so that even a big monster could hide easily. You see? It is hard
to prove the existence of a rare monster that seems to come out only infrequently,
but equally difficult to disprove it.

Well, Loch Ness has a connection with the sea and is filled with large salmon,
so it would be *possible* to have a large fish-eating animal there, but what kind of
animal would it be? The imaginary picture is usually that of some dinosaur-snake
kind of creature. There *are* photographs of—something—in Loch Ness, but the
pictures either are blurry, or have no scale in them, so we can't figure out how big
the creature is.

We know from the study of endangered and extinct species that it is possible for the
population level of an animal species to fall lower and lower, until finally there are only

a couple of them in the world, and then none. Could the Loch Ness monster be perhaps the last of his or her kind? Perhaps a freshwater version of the long-extinct ichthyosaurs and plesiosaurs that once roamed the oceans? We cannot say definitely not, but it is extremely unlikely. If it were a surface-feeding animal, we would see it more often, and if it fed at the depths, we would not see it at all. The same objection could be raised to some of the other hypotheses for the monster, including a new species of giant sea otter, a huge water snake, an enormous sea slug, and the perennial favorite, an animal completely new to science that fits into no known category.

What could have started the Loch Ness story, if there really isn't a Loch Ness monster? If we look at the sea serpent, we might be able to get a clue.

There have been stories of sea serpents from ancient times. Of course, there are marine snakes, but they're only about 6 feet long, so they're not too likely as a source for the legend. How about a giant eel? The problem here is that the longest known eel species is under 10 feet long, and they don't come to the surface either. There is an intriguing possibility, though. Eels have a larval form, like a tadpole, and this larval form is about 5 inches long, compared with a 5-foot-long adult eel. Off the coast of South America, eel larvae have been caught that are about 5 *feet* long—and an adult of this species has never been found. Now, the same size relationship doesn't necessarily follow—maybe the adult is only a little bigger, just as some frogs are about the same size as their tadpoles, but I'd certainly like to find one of those adults, wouldn't you?

How about a big regular fish? Well, most of the sea serpent stories come from sailors, who may be superstitious, but at least they know what a fish looks like. There are some very strange-looking fish, however. There's something called an "oarfish" that has an eel-like body but has a head that vaguely looks like a lion's. Very unusual. But it is only about 12 feet long, and is found in the deep ocean, not on the surface.

Whales are certainly big enough to cause you to worry about your ship, but whales look like whales, not serpents. There is one possibility that I rather like for the *real* animal that some sailors might have seen, had the bejabbers scared out of them, and then embellished with each retelling. That animal is the giant squid.

Some of you may be familiar with the squid that Mediterranean and Asian people like to eat. It is only about a foot long. There is another squid, the giant squid, specimens of which have been found which have been almost 50 feet long. There is indirect evidence that they may be about 60 feet long as a maximum size. Most of that length is in the feeding tentacles, of which there are two. At the end of the tentacle is a paddlelike swelling. Now imagine yourself to be a fisherman, in an open boat maybe 20 feet long. It is almost dark, the light is no good, you are just about ready to pull in your net, when you look over the side and see something that is a foot or two thick, maybe twice as long as your boat, with what looks, in the dark, like a head rearing out of the water. Presto! A sea serpent. In actuality, the feeding tentacle of a dead giant squid. Why dead? Because the giant squid normally live in mid-waters and never come to the surface. The only really implausible part of this story is the improbability that the dead squid would float to the surface— normally, it would be expected to sink. But, a one in a million circumstance, maybe

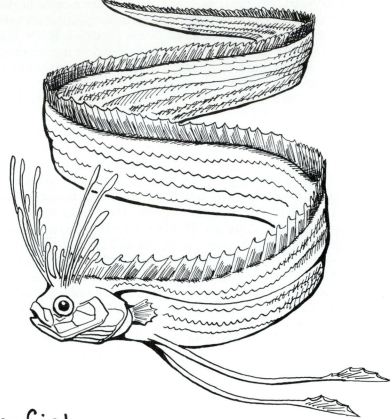

Oarfish

some buoyant digestive gases in the squid's body, and you have the sea serpent story. It takes only one imaginative, credible witness to start a myth.

Going back to Loch Ness, it is more difficult to find a candidate for the *real* animal. Giant squid are out, as are any other sort of marine animal that normally lives in deep water—they would never rise to the shallow entrance that connects Loch Ness to the ocean. We know that whales sometimes enter shallow bays and are stranded, but we need something that doesn't look like a whale, has an indefinite kind of shape, and is big enough to cause excitement. I would like to say a walrus, which would certainly be unexpected in a Scottish lake, but there are no walruses anywhere near Scotland. Loch Ness must remain an enigma, I'm afraid.

Let us now look at some other legends. The werewolf. The werewolf story started in central Europe. Hungary, Austria, Yugoslavia, places like that. The werewolf myth is one of a general class of myths that are called *transmogrification stories*; wonderful word, *transmogrification*. It means "to turn into the body of something else"; in this case, a human turns into a wolf.

Nowadays, everybody thinks that wolves are these poor, misunderstood great big puppies, but 600 years ago, wolves were greatly feared by the peasants of Eu-

rope. Their size, strength, and habit of attacking in packs made walking by oneself through the mountain forests something to be approached with fear.

Suppose that in 1450 in Transylvania there appeared a very *peculiar* stranger. His body was covered with thick hair and he walked only by night. Sometime in the middle of the night he howled, almost like a—wolf. How would the villagers explain this appearance and behavior—especially if there was a murder in the village after the stranger's arrival?

People have a tendency to explain mysterious things in terms of beliefs they already have. If we are confronted by something strange today, we tend to look for a scientific explanation. Heretofore, people looked for alternative ways of explanation.

At the time of the werewolf legend, most people of Central Europe were Roman Catholics, and one of the central beliefs of Roman Catholicism is *transubstantiation*, the belief that the bread and wine served during the mass physically change into the substance of the body and blood of Christ. To a nonscientific person who had that belief in 1450, it would not seem terribly unreasonable to believe that a person who *looked* like a wolf and *acted* like a wolf had somehow *turned into* a wolf. Such a belief, which we would now call superstitious, would not have been silly at the time, because there was no other explanation possible.

How would we explain this mysterious stranger today? Suppose there were born in a village a boy whose body was covered with what appeared to be fur. He grows up. This child would probably not have an easy time of it. Children would tease him or try to hurt him—children are often cruel to those who are different. This fur-covered child might grow up to have a nasty attitude toward life. He might be driven from his village and forced to wander. With his disposition, he might turn to a life of violence. What do you think people would say if they saw a man commit a murder, a man who had fur—like a wolf! Surely he was a man who had turned *into* a wolf!

Thus you have the beginnings of your werewolf legend. There *are* certain very rare inherited conditions that cause the body to be entirely covered with dense hair. A child born with this condition in medieval central Europe would be regarded with superstitious awe, and possibly terror. By the rumor process, it is likely that long, sharp teeth, and perhaps a tail, would be added to the description.

Later, the legend would pick up its trappings—the immortality of the werewolf, the ability to be killed with a silver bullet, and so on. Each telling of the story would add a new twist. Naturally, when you kill a werewolf, the body turns back into that of a human. The story is thus not falsifiable. This is why we do not have any were-wolf corpses to confirm the story—they have all turned back into their human form.

Come now with me from Europe to the island of Haiti, where the legend of the zombie probably originated. A zombie is an undead. He has been declared dead,

buried, and rises again from the grave. He walks with a strange gait, follows the direction of a witch doctor, and has fearsome strength. Zombies do exist; they have been buried alive, but they are not supernatural.

In the last several years, the secret of the zombie may have been revealed, although the hypothesis is still very controversial. On the island of Haiti there is a religion called *voodoo*, which originated in Africa and is a mixture of Roman Catholicism and worship of animal and devil spirits. The priests of voodoo are called *witch doctors*, and indeed they are physicians, much of their business being the use of folk medicines to cure diseases. Some of these folk medicines have been found as effective as western laboratory-derived medicines, and thus scientists found their way to Haiti.

While investigating these folk medicines, it was found that the witch doctors had discovered that if you gave a person a certain dosage of a drug derived from a native plant, that person's metabolic rate, respiration rate, and heartbeat could be so reduced that the person could actually be buried for hours, possibly even days, and then be dug up and recover. If the dosage were a little too strong, however, there would be brain damage and loss of muscle control, so the person would walk with a stagger—he would walk like a zombie, in other words.

Owing partly to the drug's action, and partly to hypnotic suggestion by the witch doctor, the zombie's willpower was reduced to nothing, and the person became virtually a slave to the witch doctor. And this, then, was essentially the motive of the witch doctor—to acquire a slave. A strong, young man would come to the witch doctor for some complaint, a cold perhaps, and the witch doctor would convert him into a zombie slave.

So the zombie is not really an undead, but as you can see, there is a possible natural explanation, embellished by the rumor process, that can explain to us how the story got started.

Finally, there is my favorite myth, the legend of the *(coughs modestly)* vampire. This is one of the more pervasive myths we know, and it has survived even to this day in innumerable horror movies. What are vampires? They are capable of living forever, but they can be killed. They must sleep on the soil of their native land, and they have the power of transmogrification. Sometimes they change into dogs, other times into bats. They cast no reflection in a mirror and are repelled by the sign of the cross. They are agents of the devil and retain their youth by drinking the blood of virgins.

Our vampire myth, like the werewolf story, originated in central Europe and represents a fascinating fusion of science, religion, and history. The belief that drinking the blood of animals (or your enemies) will give you strength is an ancient and widespread one amongst hunter-warrior peoples. It is not much of a change to shift from a belief that drinking blood will make you strong, to one that blood drinking will make you young.

The idea that the blood of Christ is a restorative is very important in most of the Christian religions, and as central Europe was strongly Roman Catholic during the years when the vampire myth was starting, it is not surprising that an agent of the devil might somehow be associated in a perverse, negative, and evil way with Christ's life-giving blood. But what was the real-world inspiration for the vampire? Here we must enter a world of greater speculation than we did with the zombie and the werewolf.

Unfortunately, it is true that on occasion one of the manifestations of murder

committed by a psychopath is the drinking of blood. Even as I speak to you of this, I can see you reacting with distress. We do not like to think of such things, but they do happen. The thought is so alien to most people that we cry out for explanation. How can such a thing happen? What could possibly cause it, we ask. Nowadays, we would look for the answer in psychology. We might even look to medicine. It is known that some deficiency diseases cause a craving for the missing factor. People with calcium deficiencies can crave chalk, for example. It is not impossible to conceive of a person with some type of anemia craving blood. If that person should be of a mentally unbalanced nature, we might have a genuine, blood-drinking murderer. Fortunately, such cases are extraordinarily rare.

But what if there were such a case in, say, Transylvania, which is part of present-day Romania. Today we would seek an explanation in science, but the ignorant peasants of hundreds of years ago would look to superstition for an answer. An agent of the devil who profaned the sacred blood of Christ would provide a satisfactory explanation. And that is probably how the vampire began.

It should be noted that in Romania in the 1400s, the time of which we have been speaking, there lived a terrible ruler who was a strong Christian. He was an enthusiast of torture, and tens of thousands of Turkish soldiers suffered unspeakable things at his hands. His name was Vlad, and he was the son of another fierce king who was spoken of as "Dracul," or the "Devil." Vlad, the Dracul's son, became known as the "little devil"—or "Dracula."

So the name of a Romanian king became linked with the legend of the vampire.

And what of the vampire bat? Did it play any role in the development of the vampire legend? Almost surely not. Vampire bats do exist, and they are a serious pest of cattle in Mexico, where they hover over the back of a cow, bite it, then lap the blood as it flows. They will also attack humans. You do not want to sleep with your toes outside the covers where vampire bats are found. The vampire bat was unknown in Europe, however, at the time the vampire legend was being formed. The bat is named after the legend, instead of the other way around.

Well, what are we to make of all this? We have seen that many of our superstitions can be explained as the rumor process working on a biological fact. Does that mean that there are no such things as *real* vampires, zombies, and werewolves? Science cannot say.

He steps from behind the podium and stands center stage, in front of the coffin. The spotlight shrinks as he talks, finally being focused only on his head.

ALUCARD: Your poet Shakespeare said, "There are more things in heaven and earth, Horatio, than are dreamt of in your philosophy." I cannot tell you if these creatures of the night and of dark places exist. But perhaps some storm-wracked evening, about the hour of twelve, you will be alone, studying, in your room. You will become aware of a chill, a feathery breath of dank air. You will look up, startled, as the curtains blow away from the window you thought you had closed. There you will see hovering, outlined against the full of the moon, the *myth* of Dracula!

As he speaks his final words, he draws his arms up, with the cape draped over them, like—wings. On the final syllable of "Dracula" two red stage bombs go off at the head and foot of the coffin. When they go out, the auditorium is in darkness for a few seconds. When the houselights go up, "Dr. Alucard" is no longer to be seen.

Day 18

Reproduction

Professor Farnsworth enters dressed very casually in sneakers, khakis, and a sport shirt. He is clean-shaven, and the students start to put two and two together.

FARNSWORTH: Good morning. Some of you thought I was dead, huh? No such luck. Ol' Viktor's a great kidder. Flew back to Bucharest last night.

Well, today, we're going to talk about a topic that's a little different to talk about than the other biological subjects we've discussed. To show you why, I want to give you a little demonstration. This is a word response exercise involving organ systems. The first thing I want you to do is find a partner. Everybody sit next to someone else. There should be a maximum of one person left over after you do this. Go. (*The students shuffle around the auditorium until everyone has a partner.*)

Good. Now I want you to face your partner. I'm going to give you two words. When I say "go," the person on the left says the first word, then the partner responds by saying the second word. We'll do this with three pairs of words. Okay, the first word is "vein," and the response is "artery."

Vein...artery. Go.

THE CLASS: Vein...artery.

FARNSWORTH: Fine. That wasn't hard, was it? Okay, the next word is "kidney," and the response is "bladder." Okay, go.

THE CLASS: Kidney...bladder.

FARNSWORTH: Terrific! Okay, now for this next one. The word is "penis" and the response is "vagina." Go.

Some students can't do it at all, others break up giggling, a scattered few are angry.

FARNSWORTH: (*Showing mock surprise*) Why should it be easy to say "pancreas" and hard to say "penis"? Well, it's complex, but in part it has to do with why we don't mind having somebody watch us eat, but we mind very much having somebody watch us go to the bathroom. There are some activities that leave us very vulnerable and unprotected. Things done in privacy have a way of becoming secret or mysterious things. Mysterious things can then become scary and uncomfortable.

256

well, sex has become such and hence is difficult to talk about. It is a difficulty worth overcoming though because it is one of the most interesting, if not *the* most interesting, biological things we will talk about.

And we must conquer our embarrassment in order to learn about this process, which unfortunately is understood by too few. Lack of understanding can lead to misfortune and unnecessary fear.

The very first question we need to ask about sex is "Why?". People tend to equate *sex* with *reproduction,* but there are many other ways to reproduce without involving sex.

The simplest way to reproduce, if you're a single cell, is just to split yourself in half. If this mitotic division is in a free-living, single-celled organism, it is called *fission*.

The offspring cells of fission are identical with the original parent cell. They have the same genetic composition, and they look like the parents and function the same way.

A little more complicated than fission is **budding**. Organisms like the hydra, which is a freshwater relative of the jellyfish, do this. A bump develops on the side of the hydra. The bump gets longer and, after eventually growing tentacles like those of the adult hydra, finally drops off the parent. The new hydra is a *clone* of the old hydra; that is, it has exactly the same genetic constitution as the original. A number of organisms have the property of being able to be artificially cloned— that is, you can take a cell, any cell, from the parent and let it grow and divide on a glass plate provided with nutrients, and it will eventually grow up into a new organism. It is easier to do this with plants than with animals—a carrot, for example, is easy to clone. In animals, the lower animals are easier to clone than the higher animals. It is still an open question why organisms that can be cloned this

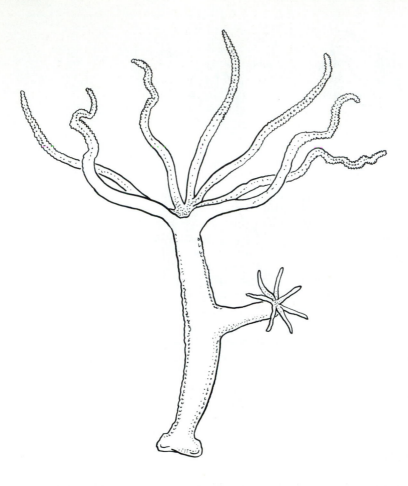

way have the property in the first place, when they don't normally clone themselves in nature. Carrots, believe it or not, are normally sexual beings, producing male and female gametes.

The kinds of reproduction we've been talking about so far are called **asexual reproduction**. Probably the strangest form of asexual reproduction is something called **parthenogenesis**. In parthenogenesis an organism that you would think is sexual because it produces eggs isn't; the eggs, instead of being fertilized with sperm, develop into an adult all by themselves. You can find this in a lot of invertebrates, but you can even find it in some lizards. For instance, there are populations of lizards in the American southwest where there are no males at all. There are endless variations on this scheme. In some cases, there is mating with a male, but the sperm doesn't actually fertilize the egg—the act of mating somehow triggers the egg to start dividing and developing.

Plants are very good at asexual, or **vegetative**, reproduction. For many plants, if you stick a cutting in water, the cut end will grow roots and the whole shebang will become a new plant. The new plant, again, will be genetically identical with the parent.

We've been talking about reproduction without sex, but you can have sex, without reproduction. There are some single-celled organisms, like bacteria, which fasten onto each other, exchange some of their genetic material, then break apart, without reproduction. This is called **conjugation**.

Okay, enough about asexual reproduction. Sexual reproduction involves the meiotic production of two kinds of cells, a sperm and an egg. The difference between the two is not absolute, but sperm are usually smaller, can move, and have little or no nutrition in the cytoplasm. Eggs are larger, have nutrients, and if they move, move passively.

Eggs and sperm are usually produced by different forms of the same organism, called *males* and *females*. Now, what is the difference between a male and a female?

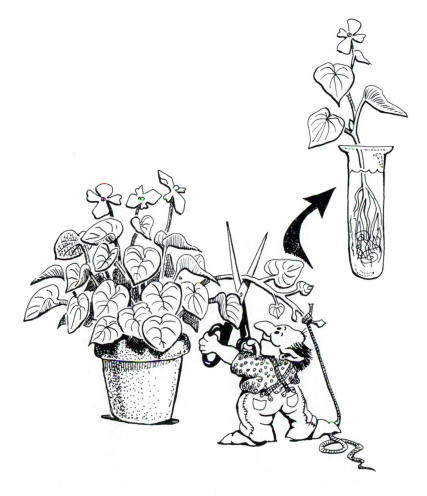

BURLY JOCK-TYPE: Males are smarter! (*Chorus of boos and catcalls from female students*)

FARNSWORTH: Ah, a very perceptive answer, sir. But this is a science class, and we do not accept unsupported hypotheses here. I just happen to have my grade book with me, so I suppose you would not mind if I read off your grade on the last exam, to provide evidence for your hypothesis? (*Raises eyebrow politely, but questioningly, as he pulls grade book out of his briefcase*)

JOCK: Well, unh, I—

FARNSWORTH: Thank you. Whatever the merits of your argument, it is not relevant to our question here, because we are talking males and females of all species, and smartness is not relevant to apple trees.

In one sense, the definitions of maleness and femaleness are circular. Males produce sperm, females produce eggs, and a thing that produces sperm is a male. Everything else is rules that have many exceptions. Sometimes males provide the nutrition for the young, as in pigeons. Sometimes the female is much bigger than the male, as with many fish. Sometimes the female is more brightly-colored than the male, as in some birds.

Where we *really* have a problem is when both eggs and sperm are produced by the body of a single individual. This is called **hermaphroditism**. If the eggs can be fertilized by sperm from the same individual, this is called *self-fertilizing hermaphroditism*. The obvious advantage of being a selfing hermaphrodite is that you never have to worry about finding a date for Saturday night, and you don't have to practice safe sex. I wasn't just kidding about Saturday night—a selfing hermaphrodite doesn't have to find another animal to mate with, and if you are a rare, or thinly distributed, animal, that can be a real plus for your success as a species.

In a **non-self-fertilizing hermaphrodite**, sperm from one animal fertilizes eggs from another—the animal's sperm can't fertilize its own eggs. A barnacle is a non-selfer.

There are some real advantages to being a non-selfing hermaphrodite. You get the benefits, which I'll explain later, of exchange of genetic material as in regular sexual reproduction, but you don't have to worry about finding a mate of the correct gender if you happen to be either rare or *sessile*, which means rooted in place. Let's say you're a barnacle. You're fastened to a rock, and you're not going anywhere. You have one neighbor, who is glued down right next to you. Well, if your species were **dioecious,** that is, had separate males and females, and your name was Albert, your one barnacle neighbor might be named George, and you'd be stuck—doomed to a life without offspring to take care of you in your old age. But by being **monoecious**, with the organs of both sexes on the same body, you can be both Albert *and* Alberta, and your neighbor is George and Georgia, so both of you get to reproduce.

There is a variant of hermaphroditism that I rather like, and that is **sequential hermaphroditism**. You start out life as one gender, then as you get older, you change into the other gender! This is usually found in small fish, and depending on the species, you start out as either a male or a female. What would be the advantages of such a bizarre arrangement? Well, here's an example. Normally, it is not a good

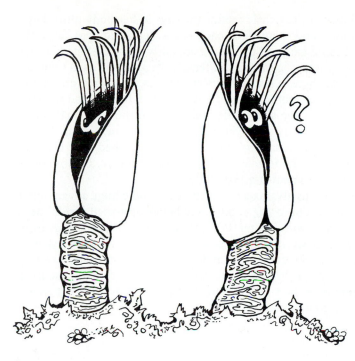

idea, genetically, to have brother-sister matings. If all the offspring are the same gender when they are young, and stick together in a school, there can't be any inbreeding between siblings. This is all speculation, though—we really don't know exactly what the advantage of sequential hermaphroditism is—it is one of nature's mysteries.

Okay, enough of hermaphrodites. Let's go to dioecious, sexually reproducing organisms. Fertilization, or **syngamy**, meaning "joining of the gametes," can happen one of two ways: external or internal fertilization. In external fertilization, both eggs and sperm are released outside the body and find each other passively. For example, a starfish female, when the time comes, simply shoots out a bunch of eggs into the seawater she lives in. The eggs float around. A male starfish does the same thing, shoots out a bunch of sperm into the water. If a starfish egg floats into an area where a male has released sperm, bingo! you've got baby starfish.

Now some advantages and disadvantages of external fertilization. A big advantage is that males and females don't have to find each other. Totally impersonal. Second, you don't waste much energy on reproduction. No courtship, no care of young, just whatever energy is required to make the eggs and sperm. That's a big plus. On the other hand, it is very chancy. If the tide goes the wrong way, all your eggs wash out to sea. You also have to be reasonably in synch with the other gender. If males shed sperm in January, and females shed eggs in June, you're out of luck. So there has to be some synchronization mechanism. A kind of indirect disadvantage is that the parents never know who their own kids are, nor do they know who the other parent is. There is thus zero incentive to evolve parental care.

With internal fertilization, the female retains the eggs in her body until they're fertilized. Now, she might pop them right back out as fertilized eggs, as with a chicken, or she might keep them until the young are fairly well developed, as with a mammal. The disadvantage is that males and females have to find each other and come close enough so that there is sperm transfer. This is a bigger problem than it might seem because in a lot of animals, the female is much bigger than the male, and she normally *eats* animals the size of the male. So somehow the male has to convince the female that he's really her lover man, and not her next meal. You see this a lot in invertebrates such as spiders; the male has developed very elaborate behaviors essentially to say to the female, "Hold on there, honey, before you do anthing drastic, let's just talk about it a little."

Getting together presents disadvantages, but there are some huge advantages to internal fertilization. The fertilized egg is infinitely better protected inside the female than it would be on its own. The mother, at least, *knows* that the eggs inside her body belong to her, and thus there is a powerful incentive for her to develop strong parental care to protect *her* genes. The male is less sure about whether a given offspring from a fertilized egg is his—while he was off hunting rabbits, she might have been having a fling with the wolf next door—but the odds are good enough to encourage at least a degree of parental care.

The result of the interplay of all these advantages and disadvantages is that females who engage in external fertilization usually release thousands or millions of eggs, because most of them will die, whereas the female practicing internal fertilization will produce eggs numbering in the single figures at any one time. There are hundreds of thousands of different kinds of animals practicing either internal or external fertilization, so we have to conclude that both methods are successful.

Plants which reproduce sexually have a variant on internal fertilization, in which the female gametes stay within the female organs of the plant but the male gametes are passively brought to the female gametes by the wind, insects, or birds. Some plants practice what is called *alternation of generations*. What happens here is that you have a diploid form of the plant. Remember, a diploid organism has 2n chromosomes. Anyway, this diploid form produces some cells by meiosis. Does anyone remember what we call the cells produced by meiosis?

STUDENT: Gametes.

FARNSWORTH: That's right, except that with some plants, like algae and fungi, these haploid cells are called *spores*. These spores settle down and grow up to be adult plants—of two different genders. Later, these haploid plants produce real haploid gametes, which fuse to form a zygote, which grows up to be a diploid plant. So here you have the same kind of plant, but during its life, it alternates between haploid and diploid forms. There are some kinds of animals, like the cnidarians—sea anemones—that do this too.

Let's talk mostly about sex in vertebrate animals for a while. In a lot of the lower

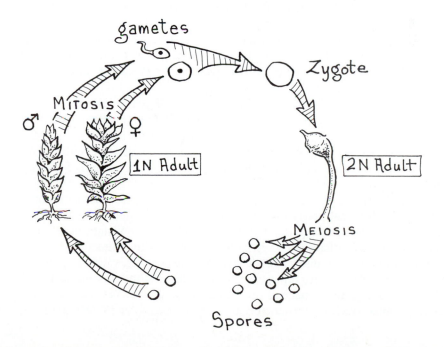

vertebrates, like many fish, there is internal fertilization, but there isn't any real degree of parental care. Males and females get together and copulate—isn't that a terrible word? I think the words used for sex ought to be really beautiful words, like *languid,* or *scintillating,* but, no, we have *copulate.* The problem for anyone with technical training is that there is another word that starts with the same three letters. Its not really a word, its a prefix—*copro-.* *Copro* means "feces," and there are all sorts of really ugly words attached to ugly phenomena—I leave them to your imagination, but they all start with *copro-,* which sounds like *copu-,* and it sort of spoils it. However, we can't really talk about toads making love, so we'll have to stick with copulation, I guess.

Anyway, after this copulation, a lot of animals just leave the fertilized eggs to their fate. In that case, there is no need for any sort of relationship between the male and the female after copulation. Once you *do* start taking care of the fertilized eggs, the rules of the game completely change. Taking care of the young 'uns might increase the chance of survival of the little tads, but for the parents, there is increased risk. If you are small, defenseless, and tasty, like a sparrow, every time you go out to get bugs for your kids, you're increasing the chance that you yourself will get picked off by a cat. Now the game starts to get interesting. In that situation, there is a real premium on getting the young out of the nest, or the den, as soon as they can be safely on their own. The second thing you would like to do is find somebody to share the labor of raising the young, hence reducing your own risk.

We humans, especially we humans in industrial societies, are spoiled. We know that rearing kids costs money,-but we don't usually think of it in terms of risk. It's very different with most animals. Even if you're a predator, like a lion, you can still get terminated if the antelope you're after hooks a horn in you. So reduction of risk in raising offspring is a *big deal* for animals.

Well, one of the best ways to cut down on your risk is to have a mate. There are mutual advantages for both genders in taking a mate. Even if the mate doesn't actually bring food to the young, the mate can look out for predators, or even fight them.

If you have internal fertilization, the female is basically stuck with the zygotes until they're ready to come out into the outside world. In tigers, males and females see each other once a year, mate, and then lead solitary lives for the rest of the year. But, you see, *tigers can get away with it.* Who is going to try to eat a tiger cub if Mom is around? The cub itself would be a formidable opponent for 90 percent of all small predators.

Not everybody weighs 700 pounds and has jaws that could crush an automobile engine block. So this small, tasty male sparrow thinks to himself—I'm anthropomorphizing, of course; sparrows don't really think about it, but natural selection picks out sparrows that behave *as if* they thought about it—he "thinks" to himself, "Jeez, I really need to get as many of my genes into the next generation as possible. Maybe if I brought the little buggers a worm or two, more of them would make it." So that's the advantage to the male of providing parental care. Yes, he does incur some extra personal risk, but presumably it is compensated for by increased survival of his young. Now it is very important to note here that these be *his* young. He is risking his own neck, and if he risks this precious commodity for some *other* male's young, he's a *biiig* loser. See, the female of a pair is in a different situation. She *knows* that those babies belong to her. The male can never be sure. The male has more to lose if the female has extra-pair copulations than the female does. If the female's mate fools around, what has she lost? So long as the mate continues to bring food to her young, or otherwise provide services, she hasn't lost anything. On the other hand, if the male's mate has an extra-pair copulation, there's a chance that those little guys in the next generation aren't going to look like him. This is why there is such a big premium on males' keeping other males away from their mates— and this is probably one of the forces that has driven aggressiveness in males—the absolutely perfect strategy for you if you are a male and can get away with it is to impregnate some *other* male's mate and let *him* run all the risks of feeding *your* offspring.

Well, when all is said and done, for many animals there is some premium on having a mate, but like almost everything else in biology, there are a variety of ways to go about things. First some vocabulary.

If you don't really have a mate that you work with to raise young, but copulate with multiple partners, you are *promiscuous*. If you have a single mate during the breeding season, you are *monogamous*, but there are two kinds of monogamy. In *serial monogamy*, you have one mate during this breeding season, but during sub-sequent breeding seasons, you have another mate, so it is one mate after another. On the other hand, if you maintain the same mate over a number of breeding sea-sons, this is *lifelong monogamy*. "Lifelong" might be stretching it a bit, but at a minimum, the partners stick together for a few years. I might add that among animals, lifelong monogamy is the rarest of all the mating systems, found only in a few species, for example, some birds and some humans.

If you have more than one mate during a breeding season, you are *polygamous*, but there are two kinds of polygamy. If you are a male and have several mates, this is called *polygyny*. If you are a female and have several mates, this is *polyandry*. Polyandry is rare, mostly found in birds, and it works like this. A number of males will build nests very close to each other. The female comes around to each male, mates with him, then lays eggs in his nest. She goes to the next male and does the same thing. After mating, the males happily sit on their nests, while *she* guards the nests of all the other males and drives away predators. In polygyny, there is usually a harem of sorts, in which one dominant male will have a number of mates and the less dominant males will end up with no mates.

What's in it for a female to be a part of a harem, rather than being monogamously mated to a single male? Well, usually the only males that can acquire a harem are the ones that can defend a large, high-quality territory that has a lot of resources available for the raising of young. This is vastly oversimplifying, but which would be better—to be the only wife of a poor guy who lives in a shack, and who gives you a '67 Valiant to drive, or to be the number two wife of a billionaire middle eastern oil sheik who keeps his wives in a palace, and gives each of them a Ferrari?

Some would prefer one, some would prefer the other. In species where you can find both monogamy and polygamy, like red-winged blackbirds, it does seem to be true that the monogamous males have inferior-quality territories.

Remember in all this, the name of the game is maximizing the number of your offspring who live to reproduce themselves. Animals don't have morals or ethics, so they can get right down to the bottom line. It is tempting to look at this biological stuff and say, "Oh, I see, this is why men are more prone to aggression than women." But there is a hugely important factor that makes the interpretation of human reproductive behavior more complex than that of other animals, and that is the factor of culture and learning. Something which might make perfectly good evolutionary sense might not make good cultural sense, and vice versa. So, it's fun to speculate about the biological basis of human behavior, but you can't just quit there—somehow you have to factor in culture.

In this discussion so far, we haven't really talked about what the real biological advantage of sexual reproduction might be. The quickest answer is "we really don't know." It certainly does provide an easy way of securing genetic diversity in a population. Meiosis produces variation in the production of gametes through recombination, crossing over, and so forth, and then just the ability to mate with different sex partners adds diversity. Why would you *want* genetic diversity? Well, let's say that as a species you are very well adapted to your present environment. Times change, and maybe a million years from now it's going to be a lot colder. Well, if there is no diversity for natural selection to act on, there will never be adaptation to the future conditions. So having the diversity provided by sex serves sort of as an insurance policy for the species against extinction if future conditions are different.

The only problem with this hypothesis is something like this—suppose you personally have a nice big insurance policy to take care of you in your old age. The only thing is that the premium is so huge every year that you don't have money to buy food. So, yes, if you live long enough, you'll be taken care of, but in the meantime you might starve to death. Same thing for diversity. If you are adapted extremely well to the present environment, any variation will make you *less* well adapted *for right now*. Natural selection acts on the here and now, so how does natural selection favor things like sexual reproduction, which are primarily useful for the future? This is a big question and is currently being debated by the biologists who study such things.

Finally for today, I want to talk about something that is pure blue-sky speculation, so realize that everything that follows here is food for thought, not gospel truth. Question: how many of you have ever been in love? I don't mean just a nice, warm, cozy feeling; I'm talking bells, walking on air, Romeo and Juliet. Raise your hands—don't be embarrassed, I've been there before. It's okay to admit it. (*A large number of students rather reluctantly raise their hands.*)

Thank you. Wonderful feeling, isn't it? But notice this remarkable thing. If you listen to somebody describing the feelings that come with romantic love, you find that the description is remarkably like that of somebody who uses uppers—amphetamines—describing the effect of the drug. Amphetamines are synthetic drugs that chemically resemble the natural hormone adrenaline, or epinephrine, which is released by the adrenal gland when you are frightened or alarmed.

Now another question—how many of you have had a love affair go sour, or have gotten unexpectedly dumped? Go ahead—I suspect it will be a lot of people. Raise hands, please. (*Again, a large number of hands*)

Terrible feeling, isn't it? Worst feeling in the world. Your eating habits change. You eat either too much or too little. You have a specific craving for chocolate. You have insomnia and start waking up at four in the morning. (*Heads nod all over the auditorium.*) Another remarkable coincidence. These are very much the symptoms of somebody going through withdrawal after quitting amphetamines.

Okay, remember, this is just speculation, but maybe romantic love—or the physical feelings associated with romantic love—is the by-product of a hormone, so far unknown to science, which is released by the trigger of romantic feelings for someone. Not a sex hormone—that's different. Maybe this hormone X is like the endorphins, which are opiumlike substances released by the brain that have the effect of reducing pain. So, when you're in love, you produce the hormone and experience the euphoria. Your lover leaves, and you stop producing the hormone; then you go through withdrawal. (*This startling idea starts the students buzzing.*)

Well, this is just an idea. Maybe when some of you get out of this place, you might want to study it. We do know that the physical aftermath of a busted romance is real, and often serious. People can get so depressed that they kill themselves. It's something worthy of study.

Okay, we've gone from amoebas to romantic love, encompassing the whole of reproductive biology in one lecture. Needless to say, we've only covered the tiniest tip of an enormous iceberg. What I hope you got out of this lecture was a feeling for the complexity of reproduction, and the many, many alternate pathways toward the same destination—the passage of genes into the next generation. Have a good weekend. (*He exits.*)

Day 19

Animal Coloration and Patterning

Professor Farnsworth enters wearing what could only be described charitably as a "colorful" outfit—Hawaiian aloha shirt in purples, pinks, black, and chartreuse; green-and-red plaid pants; and a lime green, broad-brimmed "pimp" hat, decorated with a long peacock feather. As he comes on stage, he pirouettes around so everyone can see.

FARNSWORTH: How do you like it—nice, huh? You don't have to worry about matching colors if you have *all* colors, right? Well, that's what we're going to talk about today. Color and pattern in animals.

The surface color or pattern of an animal can serve one or more of three functions. First, the surface can provide *camouflage*, to hide the animal from predator or prey. Second, color or pattern can function in *communication*. Finally, an animal's color can be significant in its *energy budget*.

Let's look at camouflage first. Naturally, if you are trying to hide from something, you'll want to know something about its sensory capability. If it has no ears, you don't have to worry about making noise. If it has no sense of smell, you don't have to worry about staying downwind. If it has no color vision, you don't really have to worry about what color you are, but you might well have to worry about what shade of gray you are.

There are some general principles of camouflage you can follow if you are confronted with an enemy that has good vision, and maybe even color vision, like all birds, a few mammals, some reptiles, and lots of fish.

The first rule is *background matching*. It doesn't matter what color you are if your color and pattern blend with the background. Let me show you.

He walks to the back of the stage, where he has arranged a projection screen that reaches all the way to the ground. He steps in front of the screen and pushes the remote control button for the first color slide. It is a photo taken inside a New York disco, illuminated with colored strobe lights. The dancing figures are clad in all sorts of outlandishly colored outfits. Professor Farnsworth blends right into the images projected on the screen.

FARNSWORTH: You see, I am rather brightly colored, but so is my background. If, however, I now change my color—

He pulls off his shirt, under which is an army camouflage olive drab and green T-shirt. His pants come off to reveal camouflage bush trousers. He takes his shoes off, and he is wearing camouflage stockings. He quickly reaches in his pocket and smears camouflage greasepaint over his face.

FARNSWORTH: I now become quite conspicuous against my background. Clearly, *camouflage* is a relative and situational term. If I now change my background (*changes slide to a picture of a jungle*), I blend right in.

Natural backgrounds are rarely solid-colored, so if you are going to have a background match, unless you live against a sand background or one of the other rather unusual uniformly colored and patterned backgrounds, it will pay to have a mottled or speckled pattern. Toads follow this scheme, and so do many moths.

In addition to a kind of generalized background matching, many smaller prey-type species, especially insects, will actually look like something else. There are many insects that look like sticks, leaves, or bark. Obviously, if you are going to play this game, you must be good at holding still.

A thing to remember in this whole discussion is that camouflage doesn't provide *complete* protection against the vision of the thing you're hiding from. It does, however, provide protection against the quick glance, and often that's all you need.

Also, even though I've said it before, I'll repeat it again. When I say about an animal, "What you want to do if you want to hide against a background of pink flamingos is color yourself pink," I'm using a figure of speech. The animal doesn't sit down, think about it, then color itself pink. Natural selection picks out the pinkish animals, but it gets cumbersome to talk about natural selection all the time. I think you get the drift by now.

All right, another strategy you can follow besides background matching and looking like something else is to break up your outline. Stripes or big blotches are good for this. The idea is that you don't want to have a clear outline of your animal shape. Better that you look like a couple of rocks than one fish. There is a principle of visual psychology that plays an important role in concealing your contour. It is called the *law of good form*, and it essentially says that once the eye starts following a line, it wants to continue along that line rather than sharply diverge. For example (*he puts a drawing on the overhead projector*), here are some lines I've drawn. I'm telling you line A is made of two lines. Does B show these two lines? No. The two lines in C are the actual lines that make up A. See what happened? You had a straight line here and a curve, and your eye wanted to continue straight on the straight line, and along the curve of the curve.

Suppose we have an animal that has a pattern on its body, and the pattern continues on the background. The eye will tend to link the two patterns, and the animal will merge into the background.

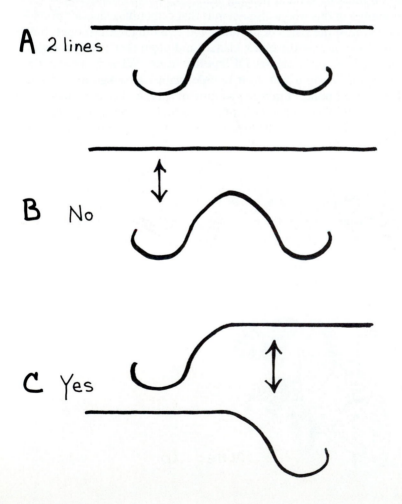

Okay, let's say I am a small, tasty animal and I don't want anyone to know I'm an animal. Already I've got a problem. How many rocks have arms and legs? (*He spread-eagles and extends his arms.*) My *limbs* are going to give me away. So, clearly, when I'm at rest, I'm going to pull my arms and legs up to my body. I can paint myself in such a way that the pattern on my body joins the pattern on my limbs when they're drawn in. This is called the *joining effect*, and it is seen on a lot of frogs. Now, do we have anyone here who plays varsity football? Ah, good. Could you tell me why you put dark greasepaint around your eyes before a game?

STUDENT: To cut down the sun's glare.

FARNSWORTH: Ah, well, that is always possible, but there are a couple of other reasons why you might want to have the region around the eyes dark. Eye movements can give a very nice clue as to which direction you are going to jump before you jump, and that is information very useful to an opponent. Putting the eyes in shadow makes it less easy to read eye movements. But consider a poor defenseless frog. It's done a good job of matching its background, and it has a joined pattern on its limbs. But there's one

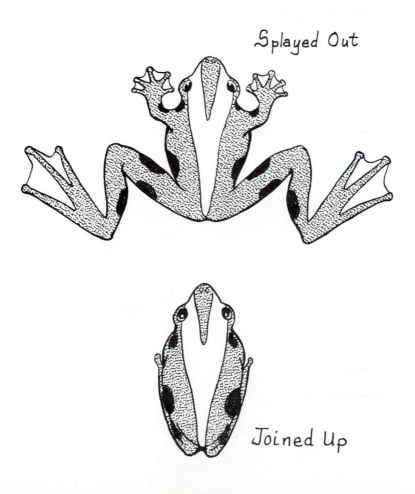

Splayed Out

Joined Up

thing it carries around with it that provides a dead giveaway that it's an animal. How many perfectly round things are there around a pond? Not many, and most of them are eyes. The roundness of the eyes is a dead giveaway. So what you do is run a stripe through your eyes. You see eye stripes mostly in defenseless, heavily preyed upon animals like frogs and small birds. Okay, could I have the spotlight on now. (*A bright spotlight goes on, sharply illuminating him.*)

You might have the best camouflage job in the world, but if you operate out in daylight, you have a real problem. Your shadow. Your shadow perfectly outlines your shape. This is more of a problem for some animals than for others, primarily tall ones, but even snakes cast a shadow, and perfect camouflage can be totally blown because of shadow.

Options are somewhat limited here. You can be active in shadowy areas of vegetation. For instance, if you're in the shadow of a tree, you will have no shadow yourself. You can have a low profile, to reduce the length of your shadow. Some insects have a very nice solution. They have wide, sloped flaps sticking out of their sides, so the shadow is diffuse, rather than sharp.

There's another aspect of being in strong light that you have to think about, too. Suppose you are a uniform color all around your body and that color is a perfect match

Countershading

for your background. Let's say you're something like an antelope. So you go out in the bright sunlight. Does it *look* like you're uniformly colored now? No, because your belly is in shadow and will look darker than the rest of you. So your belly will now contrast with the background, revealing the contour of your lower edge. Now, there's a phenomenon that you see in quite a few animals that may be a solution to this problem. Animals which are **countershaded** actually have their bellies a lighter color than their dorsal side, or back. In shadow, such an animal would be obviously lighter on the underside, but in strong light, the color would appear to be uniform over the whole body. There's no question that many animals have this countershading—birds, mammals, fish, to name a few—but the lightening of the belly shadow is, I'm sure, not the only explanation for the countershading. For example, a lot of mice are countershaded, but when they're at rest, their bellies are right on the ground and you can't even *see* their underside. So the whole story of countershading isn't known.

Okay, I think you get the general idea about camouflage. Now let's talk about exactly the opposite kind of situation—you *want* to be seen; you want to communicate something to another animal.

To make sense of the use of color and pattern in communication, I'm going to have to jump a little ahead of myself and talk about a particular principle of animal behavior. I need somebody who has a boyfriend or a girlfriend—raise hands. Okay, with the tie and sport jacket, I assume it's a girlfriend we're talking about? Okay, offhand, could you tell me how many nose hairs you can normally see in her left nostril?

SPORT JACKET: (*Embarrassed laughter*) I never paid any attention.

FARNSWORTH: Quite reasonably so. But that information is there for you to obtain, true? Now another question, sir. What color are her eyes?

THE STUDENT: Blue.

FARNSWORTH: Ah, blue. That you did notice. So both the number of nose hairs and the color of the eyes are there for you to see, but you pay attention to only one of these stimuli. We call a stimulus that conveys meaning to the recipient of the stimulus a *sign stimulus*. So blue eyes are a sign stimulus, but nose hairs are not.

We usually see bright colors or distinctive patterns used as sign stimuli which convey some kind of message. For example, take herring gulls, which are a kind of gray-and-white seagull. The body and wings are very inconspicuous, but the bill is a bright yellow, and at the tip of the bill is a bright-red spot. Why have this spot? Well, think about the situation of a baby gull chick. There it is on a nest on the ground, essentially defenseless. Along comes this great big animal, big enough to eat it. How does it know that this animal is Mom, and not a cat? Now, remember, much as I like birds, when God was handing out the brains to all the animals, the birds stopped off for a beer and missed out. What we want is a method of identification that requires very little in the way of either sensory ability or brainpower. Can the chick see colors? Yup. So what we do is color-code the parent with a red

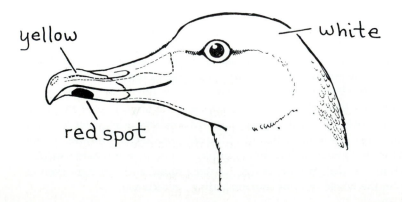

spot on the end of a yellow stick. The important thing about this is that it now doesn't matter what the rest of the gull looks like. The chick doesn't have to notice whether the animal has feathers, or the outline of a gull, or anything else. So long as there isn't anything which eats gull chicks which *also* has a yellow stick with a red spot protruding out of it, the chick can safely start begging for food rather than hiding, and can continue for as long as it sees that stick and spot.

Now we humans, being perverse, brainy creatures, can easily fool this system. We can walk up to a gull chick with a yellow-painted, red-spotted stick held out in front of us, and so long as we make gull-type noises too, the chick will start begging food from us. But how many cats are going to be able to figure this out and start walking around with red-spotted sticks? In nature, a very simple system of color and pattern sign stimuli can provide *identification*.

Identifications of various sorts are very important in the animal world. Parents have to identify their children so they don't eat them. Children have to identify their parents so they can persuade the parents to feed them. Male animals have to identify females, and vice versa. Animals of one species have to identify others of the same species when they are looking for a mate. Remind me to tell you the story about the zookeeper, the drunk, and the 800-pound gorilla someday—but I don't have time now.

Changing gears, how many of you have ever faced an enraged male mandrill, which, for those of you not familiar with it, is a very large, very ugly-tempered African monkey? Well, I have, under circumstances I would just as soon forget about, but the thing that was so distinctive about the experience was how *big* his mouth looked. I mean, that sucker was *all* teeth. And the apparent size of his mouth provides a graphic demonstration of the use of color for **intimidation**, because a mandrill's mouth is rimmed with red tissue. It is a very curious property of visual systems that things which are red look much bigger than objects showing other colors, and it is especially true that red things look big when contrasted with blue. So if you want to emphasize the size of your mouth and your teeth, you paint the outside of your mouth red.

You find that a lot of animals have red genital areas, which presumably makes them sexually more alluring, possibly by emphasizing the size of the genitalia.

Another use of bright colors is in *warning*. A color which is used to warn another animal about something is called an **aposematic color**. Aposematic colors are generally some variant of yellow. Have you ever thought about why school buses are yellow? Experiments show that yellow is the color first picked out of a group of colors by the eye. Yellow is literally eye-catching. So, if you want to catch something's attention, yellow, or possibly a combination of yellow and something else, is a good color scheme to adopt.

Think about it. What color is a yellow jacket wasp, eh? What color is a bee? Yellowish-brown. What color is the deadly coral snake? Yellow, red, and black. Most of the time, where yellow is used as an aposematic color, the thing that is warned about is the presence of a poison or venom.

It might occur to you, "Why, if you are deadly poisonous, would you need to warn somebody? If you're molested, just bite or sting the molester." The problem

is, by the time the poison works, you could already be eaten or damaged. Better to prevent the attack in the first place, and that presumably was the selective advantage in evolving aposematic colors to go along with the evolution of a venom defense.

"But," you might say, "if yellow animals are avoided by predators, why not become yellow even if you *aren't* venomous." Cheat, in other words. Well, this does occur, but in biology we call it **mimicry**, Tasty, innocuous animals looking like venomous ones. Well, my mother once told me "cheaters never prosper." That is true in Biology 100, but in the animal world a few animals have evolved as successful cheaters. The reason that mimicry is not more widespread seems to be associated with the observation that if *many* animals were mimics, the value of the bright color or pattern would be lost. If *everybody* is yellow, there would be no way to pick out the venomous ones, and most of the yellow animals you would eat would, in fact, be okay to molest and eat.

There are some animals that use a two-level strategy of color and pattern for defense. For example, there are some moths which have the outer pair of wings well-camouflaged so that they match the bark the animal normally rests on. The underwings on the other hand are a bright red. When the animal is at rest, the outer, camouflaged wings are folded over the bright ones, so the animal is well-camouflaged. The primary predators on these moths are birds, and, again, much

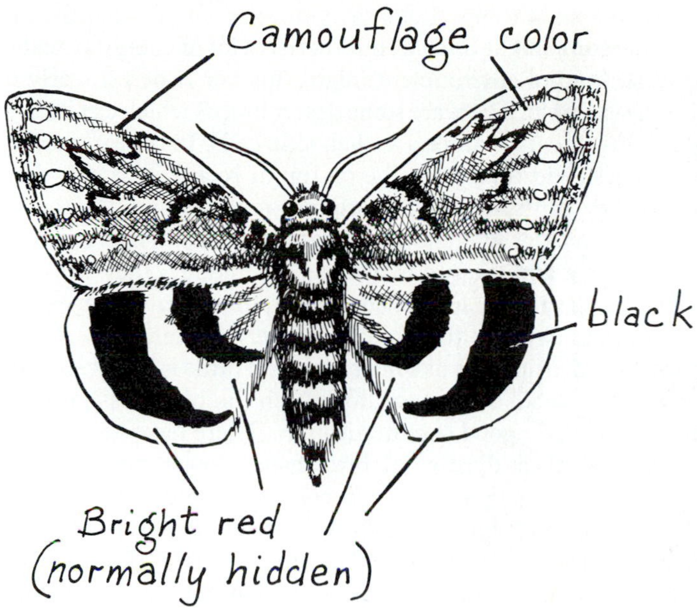

Camouflage color

black

Bright red
(normally hidden)

as I am a devoted fan of birds, sometimes as hunters they're not too swift. If a bird sees through the camouflage and attacks the moth, but misses on the first bill-hit, what the moth immediately does is FLAASH those bright underwings at the bird, which momentarily startles it; in the meantime, the moth flings itself off the tree. So if the moth is not totally protected by the camouflage, it has a second line of defense.

Okay, camouflage, communication, now we come to the final consideration of surface color, and that is energy balance. Let me ask you something. On a hot summer day, if you leave your car out in the sun, which is going to absorb more sunlight, a white car or a black car?

STUDENT: A black car.

FARNSWORTH: Good. Now a trick question. You leave both cars out in the lot and wait until night falls. Which car will *lose* more heat at night?

SAME STUDENT: The black car as well.

FARNSWORTH: Seems reasonable, but that's wrong. They'll lose heat at the same rate. During the daytime, you are absorbing visible, ultraviolet, and infrared energy from the sun. At night, you are radiating only invisible infrared energy—after all, you don't glow in the dark. Now, the thing is, the black car, because of the black pigment, absorbs a lot more ultraviolet and visible light energy than the white car, which reflects both. But at night, neither of the cars gives off any ultraviolet or visible light, and both of them radiate infrared heat radiation at the same rate. So at the end of the day, the black car is hotter than the white one and will take longer to come down to air temperature after dark.

Think about this now in animal terms. Let's say you're an animal that lives in a place where the air temperature might be cold, but there is a lot of energy available from solar radiation. What kind of environment might this be? A desert, early in the morning, would be one example. There are some desert lizards which can change their surface color by moving pigments around in their skin cells. In the early morning, these guys are very dark, and they sunbathe on top of rocks. This gets their body temperature up so they can start moving around and foraging. As the air temperature gets warmer, their color lightens up so that they can still operate, even when the air temperature is hot and the sun is blazing. At midday, they will seek shelter from the sun. By having variable surface coloration, these lizards can extend the amount of time during the day that they can be actively foraging.

The situation in warm-blooded animals like mammals and birds is less clear. You certainly do have black warm-bloods like crows, for which the blackness provides neither camouflage nor particularly good communication. There have been experiments which have shown that black-dyed white birds have a lower metabolic rate than white birds when they're both placed in cold conditions and exposed to an artificial sun. As it turns out, black birds usually don't have to worry too much about predation, either because they're protected by their habit of flocking, like blackbirds, or by their size, like vultures. So it is possible that natural selection has picked blackness rather than brownness or grayness in these animals for energy reasons, since the predation problem is resolved in other ways.

Whiteness in animals is tougher to explain. In some cases, it is clearly camouflage, for example, polar bears hiding from their prey—seals. In other cases, the explanation for whiteness is problematical. Why should many gulls and terns be almost white? Well, this is just speculation, but a lot of these white birds have very thick layers of feathers, which provide good insulation. So maybe these birds are taking care of their energy problem by good insulation, but if they're out in an area where there is a lot of sunlight, they're risking heat stress, so they go white to reflect the sun's rays. As I said, the jury is still out on the question of white animals, though.

Finally, we come to the interesting question of human skin coloration. How come we have black people and brownish-pink people? We don't really have "white" people, by the way. Stick a white person's hand on a piece of typing paper, and you'll see what I mean. I'll call them "white," though, to avoid confusion.

Our question is really, "What are the selective advantages of white and black coloration?" I can't give you any answers, but I can give you a couple of hypotheses.

Most of the fossil evidence suggests a tropical African origin for humans. Tropical Africa has a lot of different kinds of habitat, but two of the dominant ones are tropical forest and open plains, or savannah. If humans had evolved originally in tropical forests, places like where the mountain gorillas are found now, a very dark skin color might have been advantageous for camouflage. Today's Ituri people, who live in deep forest, and are very darkly colored, are almost invisible in their tribal areas. When humans then moved out of the forest, a dark color would no longer have been advantageous and a lighter color might have been selected. That's one hypothesis.

Another idea has to do with energy balance. We know that even in tropical areas, especially in the highlands, it can get cold at night. A black animal that could sunbathe early in the morning might be able to extend its foraging day. There is some support for this idea from the gibbons, the long-armed great apes of Asia. There are two color forms in the gibbons—a medium-gray one, and one which is almost black. If you go to a zoo where they have both kinds, and you observe them at sunrise, the gray one is off in a corner, but the black one sunbathes with its back to the sun, then goes off hunting bananas or whatever. So maybe early humans were highland animals and were black for energy reasons. When clothing was developed, the black coloration was no longer significant. Hypothesis 2.

Idea number 3 has to do with protection against ultraviolet light. At the equator, the sunlight is very intense, and white people exposed to the sun have a greatly increased risk of skin cancer. The sunlight penetrates right through the outer layer of cells to the cancer-susceptible dermal cells underneath. The pigment which produces blackness—melanin—acts as a uv blocker, so if early humans were creatures of the tropical savannahs, maybe they evolved a black skin coloration in response to selection by skin cancers.

The final notion that I'm aware of has to do with vitamin D. Vitamin D is made in skin cells by the action of uv light on cholesterol. Now, if you don't have enough vitamin D, you're in trouble with a D deficiency, but if you have *too much* D, you're in trouble, too, with a vitaminosis. So maybe if early humans were on the savannahs, they evolved blackness to cut down on the vitamin D production. As they later expanded to temperate and arctic areas, the melanin was no longer an advantage but a disadvantage. We do know that black people who live in areas like Scandinavia, where the sun's intensity is low, need vitamin D supplements.

So, which hypothesis is right? I don't know—maybe none of them or all of them. We'd really have to know exactly what the environmental conditions were when skin color was being evolved.

Okay, let's wrap this up. As we've seen, color and pattern play an important role in the life of an animal. Whether a color serves for communication, camouflage, or energy depends in large measure on what the most important and immediate survival problem is. If the animal doesn't have to worry too much about preying or being preyed upon, it can afford to start looking at color for communication. If it has both communication and predation covered, energy might be the factor that determines final color. In some cases, a combination of factors may be in operation. As for me (*he reaches for the pile of discarded loud clothing and starts to put the pieces on over his camo outfit*), I guess you'd have to say I have more in common with the peacock than the turkey. See you next time. (*He finishes dressing and exits.*)

Day 20

Ecology

Professor Farnsworth enters wearing a heavy wool British herringbone suit.

FARNSWORTH: Good morning. Can someone tell me what an n-dimensional hypervolume is? No? Well, by the end of the hour, you'll know what one is, or one of us will be dead.

Today, we're going to talk about a topic which is in the news almost every day, but which almost no one understands. ***Ecology***. Think about what the word means. *Eco-*. *Eco*nomics, which means the "numbers of the house." *Eco*logy, "the study of the house." In this case *house* means "surroundings." Ecology is a devilish subject to introduce because it is not about any *thing*, but is about the relationships between things.

In some senses, ecology is a little tougher to approach than some of the other fields we've looked at, like genetics, because it covers so many different kinds of topics. All of them are related in a general way, but we'll do more skipping around than usual.

Okay, if you're going to study the surroundings of an animal or plant, you have to have a fairly precise way of describing those surroundings. The easiest way to consider the surroundings of an organism is to describe its ***habitat***. You can think of a habitat as a kind of street address. A swamp is a habitat. A grassland is a habitat. A redwood forest is a habitat. Animals and plants tend to have characteristic habitats. You don't find frogs wandering around in sand dunes. There are, however, some organisms that can live in a wide variety of habitats. Humans, for example, can live almost anywhere. Other organisms, however, have very restrictive habitats. The head louse, which lives in people's hair, never lives south of the neck. The body louse, which lives in body hair, never lives north of the neck. Usually, the more specialized the organism, the more restrictive the habitat, but that's kind of a chicken and egg thing—does the habitat cause the critter to be specialized, or because it's specialized, can it live only in a narrow habitat?

Well, there are all sorts of defined habitats, and ecologists can have fun making lists of habitats and the different kinds of organisms that live there, but there is another way of expressing the relationship of an organism to its surroundings, and that is the ***niche***.

280

If a habitat is a street address, a niche is like a job description. An organism has a huge number of individual relationships with other organisms in its vicinity. If it is a predator, it has a relationship with the prey species—if the prey species should go extinct, the predator is in deep trouble. If it is a bird, and nests in holes in trees, it has a relationship with the woodpeckers that made the hole in the first place and with all the other hole-nesting birds it is competing with for access to these holes. It has a relationship to its parasites. It has a relationship to the grass it eats, if it's a grazer. Most of these relationships are mutual. Prey affect predators, and predators affect prey.

The sum of all the relationships in an organism's existence is its niche in life. The word *niche* in ordinary English means a "little hollow place in a wall where you can put something, like a statue," so a biological niche is the little place where an organism fits in the general scheme of things.

A niche describes an *n*-dimensional hypervolume. Remember I threatened you with that phrase at the beginning of lecture? What does it mean? We are accus-

What eats it

What it eats

The place where it nests

A Niche

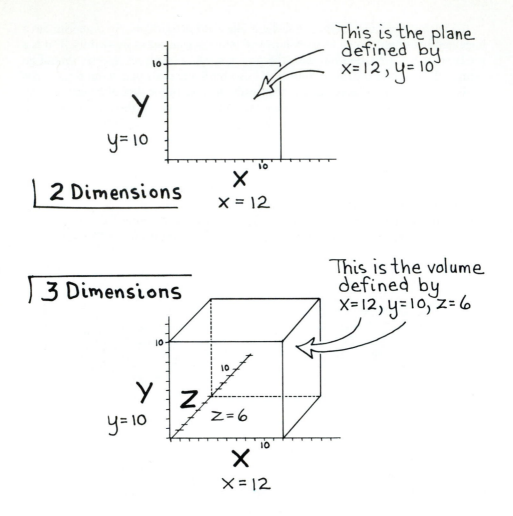

tomed to describing things usually in terms of three dimensions, length, width, and height; the *x, y,* and *z* axes on a graph. (*He places a transparency on the overhead projector.*) Well, if you have two dimensions, say *x and y,* you describe a plane of a certain *area.* A piece of typing paper has the dimension of $8\frac{1}{2} \times 11$ inches, which means it has an area of about 94 square inches. So we can describe the *area* of the paper by giving its dimensions. If we're talking about a box, now, we add a third dimension. A box $2 \times 3 \times 4$ inches would have a volume of, let's see, 24 cubic inches. So now we describe a *volume* with *three* dimensions. But suppose we had a series of five boxes, and they were all at different temperatures, because we'd heated them for different amounts of time. Could we put the temperatures of the boxes on a scale of some sort? Yes, high temperatures at one end, low temperatures at the other end. Would that temperature scale now constitute a *dimension* of the boxes? Yes. Here now is a box that has a certain length, width, height, and temperature. Well, if two dimensions describe a plane, and three dimensions describe a volume,

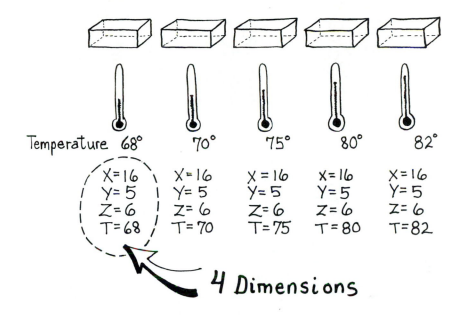

Temperature 68°　　　70°　　　75°　　　80°　　　82°

X=16　X=16　X = 16　X=16　X=16
Y=5　Y= 5　Y= 5　Y= 5　Y= 5
Z=6　Z= 6　Z= 6　Z= 6　Z= 6
T=68　T= 70　T=75　T = 80　T=82

4 Dimensions

what do more than three dimensions describe? A *hyper*volume! If the hypervolume has an undefined number of dimensions, then it is an ***n-dimensional hypervolume***, and that's what a niche is.

What are the ecological dimensions? There are a huge number. Does the animal eat big animals or small animals, or something in between? Does it live in hot areas or cold areas? Is it eaten by big hawks or small hawks? Does it breed in the spring or the fall? By the time you add up these dimensions, you have described the organism's interactions with its surroundings and with other organisms.

Why do ecologists pay such attention to the niche? Because there are a number of ideas about the survival of species that revolve around the niche idea. One idea is that you can have only one species per niche. If there are several species competing for the resources characterized by a particular niche, one of them will drive the other to extinction. This is called the ***competitive exclusion principle***, and it is controversial and not universally accepted. What it suggests is that if two species are in competition for the same resource, whether it's a limited food supply, limited number of nesting cavities, or whatever, one of them will, in the long run, displace the other. Now, competition here doesn't necessarily mean fighting in the physical sense. It could be much more subtle than that. If two toad species both breed in temporary spring ponds, and if one of them arrives at the pond earlier than the other, there won't be any room left for the other species and it won't get to breed.

Okay, so much for the niche. Let's change gears here. Ecology is a study of relationships, and there is a hierarchical arrangement of relationships. You remember hierarchical relationships from classification. You have a lot of species in a genus, a lot of genera in a family, a lot of families in an order, and so on. Well, you have a similar thing in ecology. The lowest level is the *individual* organism. You.

Me. Your dog. The flea on your dog. A tomato plant. In ecology, the individual is usually not very important. Why? Because what happens to an individual organism rarely influences what happens to the whole species. If I die tomorrow, my friends will mourn but my death won't really make that much difference to humankind a couple of centuries from now. There is an important exception. If you have a very rare species, for which there are only a few individuals left, then the fate of an individual becomes very important.

Above the individual is the first really important ecological level of organization, and that is the ***population***. A population is all the individuals of a given species living in a defined area. We could talk about the human population of the auditorium. We could talk about the population of rye grass plants on the Quadrangle. We could talk about the population of *Staphylococcus aureus* bacteria living in your throat. These are all populations.

Above the population is the *community*, and a community is all the populations of organisms living in a defined area. We could talk about the community of this biological sciences auditorium. You're surprised that there is a community? Well, in the rug of the auditorium are millions of microscopic spiders, mites, and insects that eat the crumbs of the Twinkies you smuggle into class. There are at least two dozen populations right there under your feet. Then, this is a big class, there's bound to be—I hate to say it, but it is true—there's bound to be at least one person here with lice. (*Smiles evilly*) Maybe the person sitting next to you. Then there are the bacteria in the air and the viruses in your guts, and by the time you add everything up, you have probably hundreds of populations in our happy little community. I think you can already see the beginnings of ecological relationships. Look down at the floor now. Right now, right under your feet. Those little guys down there are *totally* dependent on you for their supply of dandruff, hair, and grass clippings you bring in on your shoes, spilled soft drinks, and other things I dare not even mention. Doesn't it make you feel good to be needed?

Well, above the community is the *ecosystem*, which is the community of an area along with the biologically important abiotic components of the area. Okay, what's a "biologically important abiotic component"? Water. Dissolved oxygen in the water of a pond. Nitrogen in the soil. Anything that is important for the maintenance of the life of the organisms within the ecosystem area.

To confuse things a little, there's another—it's not really a definition—but another concept attached to *ecosystem*. An ecosystem is the smallest self-contained (with the exception of energy) ecological unit. Remember in the fourth grade, your teacher had you make a "balanced aquarium"? It had some fish, it had some green plants, and it had some snails to eat the fish and the dead plants, and the snail feces could then be used as fertilizer by the plants. You didn't have to add fish food or anything to the system, once you started it. *But,* you had to expose the whole business to sunlight, or otherwise the system would have run out of energy, the plants would have died, and the fish would have followed. The balanced aquarium is an ecosystem. A city isn't really an ecosystem in this sense, because a city has to import food from outside the city, so it isn't self-contained.

The ecosystem is a very important part of ecological study, and it has been found that there are different components of the ecosystem. First, are the *primary producers*. These are the green plants, which take solar energy and convert it to food. An organism that eats the plants is a *primary consumer*, like a horse. An organism that eats primary consumers is a *secondary consumer*, like a fox that eats rabbits. An organism that eats secondary consumers is a *tertiary consumer*, and so forth. Now, let me see if you have this straight. Suppose you had a vegetarian, whose ship was wrecked off a south Pacific island, washed ashore, and a cannibal ate him. What would the cannibal be?

STUDENT: A secondary consumer.

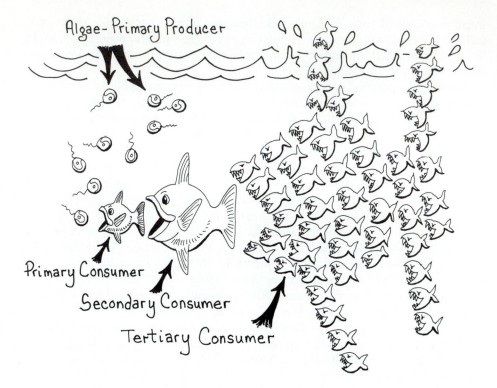

FARNSWORTH: Correct, but if the castaway was not a vegetarian, but loved seafood, the cannibal would be a tertiary consumer, or even a quaternary consumer, depending on what kind of seafood it was. Clearly this is a simplistic view—how do you classify omnivores, animals which eat anything?—but it is the key to some very useful concepts.

Now, suppose we had an acre of grassland. How many consumers would you have relative to producers, and within the consumer class, would you have more primary consumers, secondary consumers, or whatever?

STUDENT: You'd have more producers, and more primary consumers than secondaries.

FARNSWORTH: Correct. You can visualize this relationship as something called a *food pyramid*, with large numbers of individual producers, like grass plants, at the bottom and a very small number of individuals, like hawks, at the top. Our next question is "Why is this so?"

The answer has to do with the amount of energy tied up in the body of the thing that's eaten. Let's say I eat a piece of lettuce. If I burned that lettuce, literally burned it with a match, I could get a certain amount of energy out of it. I could consider it a fuel. But if I eat it instead, do I convert all the energy in that lettuce leaf into energy for me? No. First of all, there's a lot of waste. A good bit of that lettuce leaf is a polysaccharide called *cellulose*, and humans don't have the enzymes

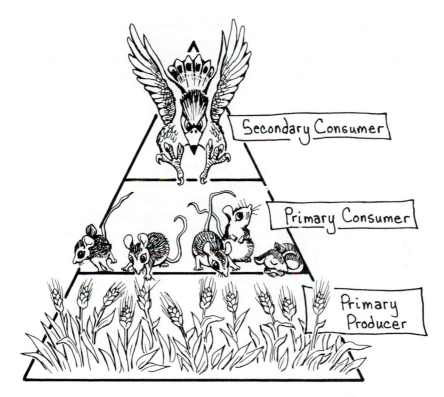

to break down cellulose. So most of the leaf goes right through me and is not available for my use. What *is* usable costs energy to process. So what we finally end up with is maybe 90 percent of the mass of the leaf lost and 10 percent retained, either to use for structural purposes or to use for energy.

The implication of this is that it takes about 100 pounds of primary producer mass to support 10 pounds of primary consumer mass. As you move from primary to secondary consumer, you have similar losses—it takes about 100 pounds of primary consumer to support 10 pounds of secondary consumer. Put another way, it takes about 1000 pounds of primary producer to support 10 pounds of secondary consumer.

I think you can see now why animals at the top of the food pyramid are so rare, like the big cats. An eagle requires a feeding area of hundreds of acres of vegetation to support the critters it eats. When humans eat animals, it requires more plant material to support the human than it would if the human ate the plants directly.

There is another implication of this food pyramid called ***biological magnification***. Let's say I have a wheat field and I spray it with a pesticide that is not immediately broken down by a vertebrate animal body, but rather stored in its fat. DDT is like that. Okay, the concentration of the pesticide on the wheat is very low, say 1 in a billion in the plant cells. Now a mouse starts eating the wheat. It stores the pesticide, rather than passing it through, so now the concentration in the cells of the mouse is increased, say to 1 in a million. We're still below toxic levels—we haven't killed the plants, and we haven't killed the mouse. So along comes a fox that eats mice. Now we have a concentration in the fox of 1 in a 100,000. This is approaching toxic levels. Finally, a hawk eats the fox, and other foxes. *Now* we have a toxic level, in the hawk, and it drops dead, or has some other serious problem like disruption of breeding.

What all this amounts to is that the higher consumers act like a natural barometer of environmental pesticide concentration, which is a very important reason, *for us,* to make sure that they are otherwise kept protected. If a hawk starts to get in trouble, we know that we're going to be in trouble soon and maybe we better see what the hell is going on. The dangers of DDT first surfaced with problems that predatory birds had, and it was a damn good thing that somebody knew something about basic ecology when they started reading about a strange thing with pelicans— all of a sudden their egg shells were so thin they were cracking under the weight of the parent. DDT was the culprit.

We kind of got sidetracked here; there is one final component of the living segment of an ecosystem, and that is the **decomposers**. A lot of the total energy of an ecosystem is tied up in dead bodies. Suppose every time a mouse died of natural causes, it just lay there forever. All that carbon, all that nitrogen, all those good energy-rich bonds going to waste. So the creepy-crawlies take over—things such as decomposing bacteria, corpse beetles, vultures. What they do is turn dead bodies into plant fertilizer. The vulture eats the dead rabbit, and when the vulture defecates, that's the most wonderful fertilizer in the world. So the physical materials in the ecosystem—the nitrogen, the carbon, the minerals—instead of being tied up forever in bodies, get recycled. Ain't nature wonderful?

Okay, the last thing we're going to mention about ecology has to do with numbers—the numbers of organisms. Sometimes you have a big population of animals in an area, and sometimes you have a small population. What is it that decides how many organisms you have in a given place?

We'll start this by considering an island. Let me just drop a small nuclear device on the island to clean it of all life-forms and give us a fresh start. Now I'm going to plant some grass, and—oh, we have to do something about the radiation. Hmmm, let's say it wasn't a nuclear device, just maybe a couple of dozen regular 12,000-pound bombs. Anyway, we start up some nice grass. This is a pretty good sized island, and we let the grass get established, and then we parachute in a couple of mice, a male and a female. Mice eat grass seeds. We start out with two mice and an unlimited supply of grass and water, and no animals that eat mice. If we come back in a while, how many mice will we have? Well, at first, the little suckers breed like crazy, so the rate of population growth increases dramatically. Once it reaches the maximum number of offspring the mice can produce, it levels off. Pretty soon, we've got an awful lot of mice on this island. We haven't increased the amount of grass seed, but the number of mice has increased, so that after a while, there's just not enough grass to go around. Some of the baby mice die of starvation, the population levels off, and the **growth rate** goes to zero. This is an important point here. We're really talking about two things: size of the population and growth rate. At

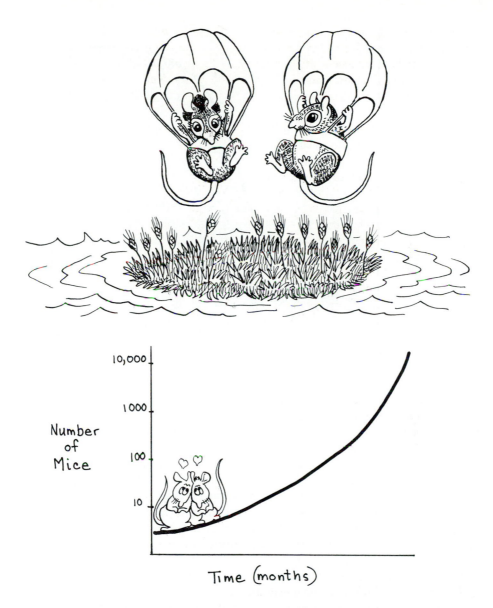

Time (months)

first the population size was small, but the growth rate was large. Then as the population size increased, the growth rate went down. Finally, the population size stabilized, and the growth rate went to zero, which means that the number of additions—births—in the population equals the mortality, or number of deaths.

This kind of S-shaped curve which describes population size is our jumping off point for considering some things that can affect a population. What it basically says is that if you start out with organisms in a new, favorable environment, the size of the population will eventually stabilize at a point that is called the *carrying capacity*

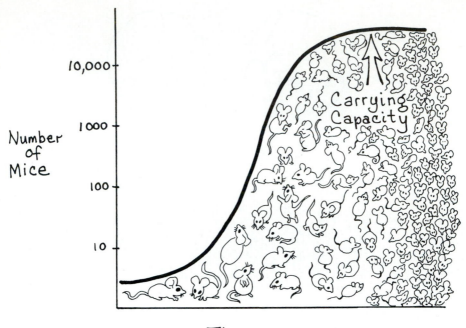

of that particular habitat. Usually, there is some one limiting resource that keeps the population from growing bigger. Let's say on our island that the limiting factor was grass seed. Could we increase the population by air dropping grass seed onto the island? Sure, and the population would increase until we hit the next limiting factor, which might be water. So we pipe water in, and the population shoots up again. Eventually, we'll reach an ultimate limiting factor, which is usually the social stress of overcrowding. The mother mice won't be able to nurse their young in private. The males will be fighting all the time instead of paying attention to the females. So we can't ever really get to the point where we have actual wall-to-wall mice, but if we artificially remove natural limiting factors, we can have a hell of a lot of mice.

But what happens if now we get bored with flying in all that grain? Suppose the mice don't pay the water bill and it gets shut off? Will the population just drop back to the point where it was before we started interfering—to the natural carrying capacity?

There is a good chance that something rather more dramatic will happen. Now you have far more mice than can be supported by natural grass seed. You've got millions of starving mice. What are they going to do? They'll eat anything they can get their teeth on. They'll cut the grass right down to the roots, and kill it. What does that do? It reduces the carrying capacity. When the population gets too large, it tends to destroy the very resources it needs for survival. This leads to a ***population crash***. The population drops way below the level originally supported; then,

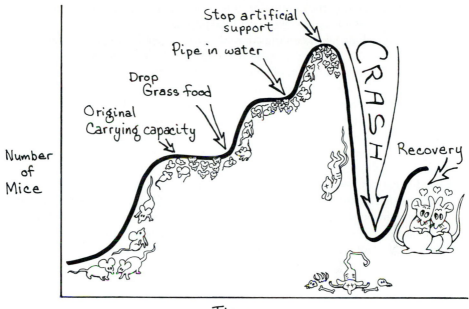

after a time, the resource base usually recovers by itself, and eventually the population returns to the original level. In the meantime, it is tough times for the survivors.

There are several observations to be made from this phenomenon. The first is that natural populations have a tendency to be cyclic. Boom and bust. The downturn may not be as dramatic as a crash, but there are normal fluctuations in population levels. What this means is that you have to be very careful in interpreting a decline in population numbers as indicating that there is something "wrong" with the population—like a pollution kill off. Populations go through normal downturns. This shows you the importance of making what are called *baseline* environmental studies. Every year, you go out and count robins on your lawn. Over 20 years, the numbers will go up and down. With this baseline, you would then be able to pick up a decline that was *higher than normal*, and then you could start getting excited about a possible environmental disaster. Otherwise, you'll be crying wolf every time there is a little downturn in a population.

The second thing you will notice is that you have to be very careful about intruding on natural population regulation mechanisms. Nobody seems to mind if rats, spiders, and snakes starve to death during the winter, but cuddly things bring out our natural inclinations to be helpful to those "less fortunate." So we'll feed the deer over winter and put out bird feeders. This artificially jacks up the population. Now, the fact of the matter is that in human terms "nature" is not a very nice place. It is *normal* to have a big chunk of the population not make it through the winter. Small animals' reproductive capacities tend to be huge as a reflection of this fact of life—even if 80 percent of the population gets knocked off during winter, that's just that much more food for the survivors, and the population makes a quick recovery.

Now we come along and keep the adorable furries going over the winter. Next spring, *too many* little guys have made it for the resources available, and you've sown the seeds for a crash. Now, of course, not every human intervention is going to have a disastrous effect—in some cases where there is an unusually bad or abnormal winter, intervention may benefit the population, but what I'm saying is that the ecological relationships of food, water, and shelter are very complex, and if you stick your nose in one place, there may well be an unanticipated ugly by-product effect someplace else. But that's why you have professional ecologists to help make these kinds of decisions.

What I hope you have learned today is that in examining the complex web of relationships that exist between organisms and their environments, there rarely are any simple right-or-wrong answers. Okay, any of you with gripes about the last exam, I'll see you in my office, but remember, you have to pass through the weapons check at the door. (*He exits, followed by a trail of students.*)

Day 21

Ethology

Professor Farnsworth enters wearing what appears to be some sort of Air Force parka. He shakes the snow off and begins.

FARNSWORTH: Good morning. Clearly, we are still in the middle of winter, but I can foresee a time when it will start to warm up, buds will come out, and we'll see the first robins of spring. By the way, does anybody know how robins find their way north during migration?

STUDENT: Instinct.

FARNSWORTH: Very good. Instinct. But what does that really mean? How does that explain anything? Is there some mysterious, unexplainable force called *instinct?* No. What we are really saying is that the robin finds its way north *without having been taught to do it.* The robin was *born* with this skill, this behavior. So when we talk about *instinctive* behavior, we really ought to speak of **innate**, or inborn, behavior.

If you are born with some kind of knowledge, that must mean that somewhere in the fertilized egg that once was the animal, the information on how to do a particular form of behavior is stored. A fertilized robin's egg contains the information on how to navigate north. Most probably, that information is stored in the DNA, along with information on building proteins, but we know so little about how behavior is stored that I won't even bother to talk about it. We *do* know quite a bit about how this inborn information operates once you have the animal hatched or born, and that will be today's topic—innate behavior.

The study of innate behavior is called *ethology*, and it is a relatively new science, maybe 50 years old or so. People have been describing innate behavior since people started watching animals, but *how* an animal did all these wonderful things like sing complex songs or build nests has been a total mystery until recently.

Before we get to specifics, there are a few—I guess we'd have to call them "precautions"—about studying animal behavior. People are very anthropocentric, which means that if we see something that looks like human behavior in an animal, we tend to think that the animal does it for the same reason that the human does, and in the same way. This is an error. As we will see, very often a very simple expla-

293

nation will account for what appears to be conscious behavior. Let me give you an example. Suppose I have a long trough, and I put a light bulb at one end of it. I dump a bunch of insects in the middle of the trough, come back in an hour, and find most of the insects at the dark end of the trough. To simplify things, let's say there's a heat filter on the light bulb, so the temperature is the same at both ends. How would you explain this? Yes, Anna?

ANNA: The insects are avoiding the light.

FARNSWORTH: Avoiding the light. Would that be sort of like, here is an insect in the middle of the trough, and he looks one way and sees the light, looks the other way and sees darkness, and decides to walk to the dark end?

ANNA: Yeah, I guess, something like that.

FARNSWORTH: Okay, but that assumes a conscious decision, a preference, a choice. It further assumes a complex nervous system that can evaluate this situation and make that kind of choice. Well, there is a principle in the study of animal behavior

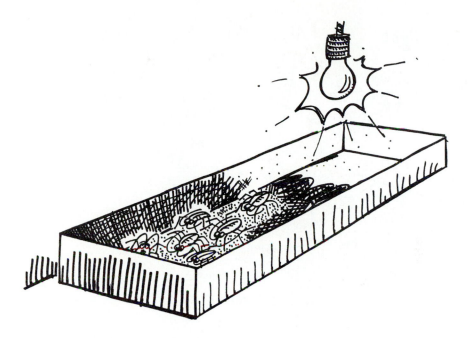

that is drawn from a more general principle of science called the ***principle of parsimony***. What it says is that if you have a choice between two explanations for something, one simple, one complex, the simple one is probably correct. In behavior, if you're trying to explain some action, you assume a simple, mechanical reason for the behavior rather than conscious thought, unless you have eliminated the simple explanation by experiment.

In the case of the insects, there is a possible simple explanation that doesn't have anything to do with conscious choice. You know how insects such as ants are always running over the ground, then turning at random? Well, suppose the distance that you travel before you turn at random is dependent on how much light there is. If it is bright, you travel a long way before turning. If it is dark, you go only a little way before turning. So what's going to happen here? Let's say the insect starts out, just by chance, going toward the light. It'll go a long way before it turns. Soon it'll head toward the dark end. When it does, it won't tend to go as far before turning. What will happen, then, is that all the insects will tend to end up at the dark end, making all these little jogs, and never getting down to the light end again. Simple mechanical explanation; no brains, or thought, or choice required.

Now this doesn't mean to say that thought or consciousness is not *possible* in animals. All it says is that you have to rule out simple mechanical explanations first.

There's a way of looking at innate behavior, a *model* as it were, that can give us a good jumping off place for looking at specific sorts of behavior. How many of you know what a jukebox is? (*Quite a few hands go up, but many don't.*) Okay, that makes me feel old, but I'll have to explain it. In the old days, when they had a lot of restaurants called *diners*, kind of cheap, greasy spoon places, they didn't have piped-

In The Trough:

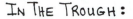

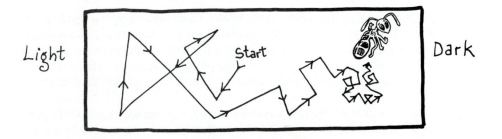

in music. They had a big machine called a *jukebox*. It was just a big record changer that held maybe 75 different records. On every table, there was a little selection box, where you put a nickel in and you could pick a record. You picked a record using two rows of buttons, one with letters, one with numbers. Every record had a number and a letter. So maybe the Stones' "I Can't Get No Satisfaction" would be *G4*. Elvis and "Heartbreak Hotel" would be *K4*. You'd punch in *K4*, and the jukebox would play "Hotel." Now let me ask you something. That record—the actual physical record of "Heartbreak Hotel"—how much information is there on that record? A thousand bits? A million bits? I don't know what the exact number is, but there is a tremendous amount of information on that record. Now, how much information do you have to feed into the jukebox to get all that information out? Only a tiny bit—a number and a letter. So a small investment in information pays off with a tremendous amount of information.

Now, before I tell you what this has to do with animals, let me ask a question. Suppose you didn't have a record inside the jukebox, but, instead, Elvis himself

and his band. Every time you put in the nickel and punched *K4*, the King would rattle off "Heartbreak Hotel." What would be the difference between that and having the record play?

STUDENT: Well, with the live performer, the performance would be a little different each time, but with the record, it would be the same every time.

FARNSWORTH: Exactly. Now suppose you couldn't see whether it was Elvis or a record in the jukebox—all you had was the speaker in front of you. Could you tell which was which if you listened only once?

SAME STUDENT: No—it would be real tough unless Elvis made so many mistakes that you would just *know* it had to be live.

FARNSWORTH: Right again. It would be very hard to tell from the *output* whether you had a thinking, conscious organism or a prerecording. Now let's get to the animal. Innate behavior is most often *stereotyped;* that is, the animal always does whatever it does in the same way, regardless of variable situations. Its song is always the same, just as the record always sounds the same. It always builds its nest in the same way. It will continue to do things the same way even if the conditions change so that a different way would be better. Stereotyped. Some kinds of behavior are very complicated, but the fact that they are stereotyped gives us a clue as to how they work.

Just as in a jukebox, where a lot of information is stored in a record, in an animal, behavior is stored in the nervous system in the form of what is called a ***fixed action pattern***. We don't know the physical basis for this, as we know the physical mechanism of a record, but the principle is the same. Singing behavior is stored as a fixed

action pattern. The bird doesn't have to think about his singing at all, no matter how complicated it is. He just "plays" his song "record." What causes him to play his song? A relatively simple information input. Just as in the jukebox, where all we had to do was feed in $K4$, to get the song out of the bird all we have to do is give him a relatively simple stimulus, something called an ***innate releasing mechanism*** (***IRM***), or *releaser*. This could be something as simple as the sight of another male bird.

There's another thing about a jukebox that can tell us something about animal behavior. When I was a kid and my parents would take me to a diner, I'd always beg a nickel off them, but I had a more sinister purpose in mind than just playing a single record. What I would try to do is punch up two input signals at *exactly the same time*. $K4$ and $C7$. Most of the time, you couldn't hit them at the exact same time, and you'd just get whatever the first one was. But if you *did* get them both at the same time, the machine would go berserk. It wasn't designed to handle it. So the record changer part would just start spinning and spinning. Eventually, it would stop randomly, and you'd get something that wasn't either of the records you punched in. I realize this doesn't sound like a helluva lot of fun as I tell it now, but when you're 9, it doesn't take much to amuse you.

"TORTURING THE JUKEBOX"

The animal works in the same way. If you get two IRMs coming in at about the same time that are conflicting, sometimes you'll get a piece of behavior that is inappropriate to either of the original stimuli. This is called *displacement* behavior. The classic example is with a bird; you have one stimulus coming in that calls for the bird to do one thing, another that calls for the opposite, so the bird will do neither. This most often happens in a fight-or-flee situation. The bird doesn't know whether to fight an interloper on its territory or run away. So it hesitates, starts to attack, backs off, starts to go away, hesitates; then it does the most amazing thing— it starts to go through the ritualized movements it would go through just before it went to sleep! So here is this invader, this Attila the Hen bird that is going to make mincemeat out of it right on its doorstep, and instead of either fighting or flying away, it starts to go to sleep! See, the system is just not built to handle tough dilemmas, so the behavior is *displaced* to something inappropriate.

There is a similar kind of phenomenon called a *redirected response*, in which the behavior is appropriate to the situation but is not directed to an appropriate object. I can give you an example of this. (*He moves over to the podium.*) Now, I like to think I'm a pretty easy guy to get along with, but if you talk to some of my colleagues after faculty meetings, they might tell you a different story. See, sometimes at faculty meetings, somebody, say, maybe somebody I've had a lot of arguments and trouble with in the past, will propose something, and what he proposes is really crazy and stupid and will be the ruin of the college if it goes through. So now it's my turn to talk, and I get more excited as I start speaking: (*He pounds moderately hard on the podium with his fist.*) "*Damn,* that's a dumb idea." (*He pounds harder.*) "*Damn,* that's the third time this year this idiot has proposed something foolish." (*He pounds still

harder.) "*Damn,* I don't know why the Dean doesn't fire this self-righteous, bigoted moron! *DAMN!*"

He pounds his fist all the way through the top of the podium with a resounding crack. The students do not know that the bottom part of the podium had been replaced with balsa wood faced with quarter-inch plywood, which makes a loud cracking sound as it splinters.

FARNSWORTH: Well, now look what I've done here. See, I was really angry with my colleague; he represented a threat to me, and in precivilized times, it would have been appropriate for me to go over and try to pound his face, but now I have a conflict. It might be very satisfying to do that, but if I do, I *know* I'll get sued, maybe spend some downtime in the slammer. Bad scene. So what I do is pound the podium instead of his face. Redirected response. You can see animals, especially birds, do this in conflict situations involving fighting. Instead of attacking the other bird, they'll viciously attack a leaf on the ground.

The thrust of this discussion is that you can see some very complex kinds of behavior in animals, but it is not necessary to assume conscious thought to explain it.

Okay, just for fun, let's briefly take a look at a couple of specific kinds of behavior. First, courtship. We've already talked about sex some; now we're talking about finding a potential mate and persuading him or her to mate with you.

The thing that colors a lot of courtship behavior is the fact that it is a hostile world out there, and you normally don't want other animals to approach you closely, unless you're pretty damned sure about their intentions. In many animals, there is a big size difference between males and females, the females usually being larger, and if the animal is a carnivore, the smaller animal has to persuade the larger that he is a lover, not dinner. We've already seen the case of the spiders; the male has to do this elaborate dance in front of the female. The same thing occurs in insects like the praying mantis and in lots of fish, in which there sometimes is a huge size difference between the male and female.

Many male birds will perform ritualized displays to stimulate the female into mating. Some of these can be spectacular. The one I particularly like is done by falcons. The female orbits around while the male climbs 1000 feet above her, then does a series of power dives, during which he can exceed 150 miles an hour. The pair then mate in flight and tumble through the air until mating is complete. Of course, we have no idea what's going on in a falcon's head, and as I pointed out,

it is a mistake to attribute human qualities to animals unless you have eliminated simpler explanations first, but *I* think that's kind of a romantic idea. Almost makes up for having to eat live mice for a diet.

Feeding behavior presents similar problems. Again, if you're a carnivore, you don't want to be eating your own children. So the young of many animals behave in a certain way, which acts as a releaser for parenting behavior, rather than eating behavior. The most obvious need for something like this is in the mouthbreeding cichlid fish. Here, the young, which are tiny, actually seek refuge in the mouth of the parent when danger threatens. The young are about the same size as the food of the parent, which is small fish. So, the young fish perform a little jig dance just out of reach of the parents. This tells the parent to open its mouth, but not swallow.

Okay, I've given you just a taste now of animal behavior. There are hundreds of other curious animal stories, and I commend them to your attention, but this is all we have time for. Next time will be our last lecture, and I want to caution you to be on time, please. See you next time. (*He exits.*)

Day 22

The Last Lecture

Sunrise, sunset;
Sunrise, sunset;
Swiftly flow the days.
Seedlings turn overnight to sunflowers,
Blossoming even as we gaze.

Sunrise, sunset;
Sunrise, sunset;
Swiftly fly the years.
One season following another,
Laden with happiness and tears.
—Sheldon Harnick
Fiddler on the Roof

It is the last day of classes. As the students file into the auditorium, they see that the stage is bare, save for a spotlit, tall stool. "Sunrise, Sunset" from Fiddler on the Roof *plays over the PA. At the stroke of the hour, Professor Farnsworth enters stage left, clad in a black turtleneck sweater, black pants, and black Wellington boots. The houselights fade to black as Farnsworth perches atop the stool.*

FARNSWORTH: This is the last lecture of the year, and the last time I will be seeing you as a class. The last time of anything is always a bittersweet occasion. It is a time for fond recollection and regret for missed opportunity. Many thousands of students have passed through this room before you. Some have gone on to brilliant careers and lives. Others have vanished in the crowd. The one thing that brought them, and you, together here was an interest in science, whether that interest was professional, or simply indicative of curiosity about something that has such a profound influence on all our lives. All semester we have been talking about the results of scientific discoveries, the marvelous things that have been found in the laboratory and field. But science is a human activity, along with art and music, and we haven't yet considered what it is like to be a scientist. This is a good time to speak of such things, because most of you are now finishing your first semester in college and some of you want to become scientists.

The freshman year is often both the best and the worst year of your life. At the end of the first semester you may be looking at grades that aren't quite what you'd like them to be, and you may be having doubts about what your future plans are. The path to becoming a scientist is a long and rocky one, and you now stand at its beginning. I'm certain that many of you have what it takes to reach the end of that road. I had a particularly disastrous first year in college, and stand before you as living proof that a bad first year does not necessarily mean death. What I would like to do today is tell you how it was for me, on this peril-filled road to science. I hope this tale will be one of encouragement to those of you who aspire to careers in science, and will give some insight to those of you following other paths about what makes a scientist tick.

I'm sure that not all scientists are born with a taste for science, but a surprising number of them show very early signs in that direction. It doesn't have anything to do with brightness or genius—just preference. When I was 5, I knew I wanted to be a scientist and a teacher. I'm sure I didn't even really know what a scientist *was*, but while the other kids wanted to wear Yankees uniforms when they grew up, I wanted a lab coat. The surprising thing to me as an adult now is that this idea was so attractive to me as a child, seeing that the Hollywood and TV images of scientists are 90 percent negative—mad scientists, evil scientists, nerd scientists, you name it. Of course, I don't really remember this, but my parents told me that when I was tiny, I would go upstairs to visit Mrs. Goldberg, who didn't have any kids, and in return for a platterful of chocolate chip cookies, I would lecture to her on the mechanisms of the motions of the planets. I must have been an insufferable little brat, but the cookies kept coming, and I saw no reason to change my ways.

My parents encouraged me in this madness with science toys. I think my Dad was disappointed when he would try to get me interested in playing catch and after about 30 seconds, I'd say, "Are we done now?" but he resigned himself to the inevitable and proceeded to supply me with an almost endless supply of wonderful science toys. The one I remember best was the chemistry set. Some of you may have had chemistry sets as kids yourself, but the old ones were much better than the ones they have now, because this was before child safety laws, which I suppose are good and necessary, but with the old sets you could do *real stuff*. You could make bombs, and though they never quite gave you enough potassium chlorate to completely blow up your kid brother, you could sure scare him out of his gourd.

Those chemistry sets came in a three-part steel box that weighed about 20 pounds, and when you unfolded it on the card table (or whatever limbo your mother consigned you to while you manufactured unthinkable and unspeakable compounds), as far as the eye could see were row after row of brown glass bottles that contained unlimited possibilities. Lined up like little transparent soldiers were the test tubes, which soon became congealed with various kinds of brown and green sludges.

Long before I had any awareness of a scientific community, there was an underground network of chemistry set owners who would meet in vacant lots after school to swap recipes. "I've got one that smells like a dead cat!" my friend Gary would offer. This would be traded for a compound Fred discovered that would make you pee blue if you ate it. Explosives and incendiaries were very big items, as were things that made quarters turn black. I have no idea why we didn't poison ourselves, or detonate half the neighborhood—I suppose that did happen some

places—but I don't remember any untoward incidents, oh, except Sam burned his left eyebrow off when he got too close to the alcohol lamp. No permanent damage, though.

There were biology kits, but they weren't particularly attractive. They consisted primarily of a series of mummified frogs and worms, a set of anatomical diagrams with all the reproductive parts removed, and some dissecting tools of a dullness guaranteed not to harm little fingers but lacking something when it came to getting through the leathery skin of those sad, gray frogs.

But the finest toy, one that has a special place in that portion of heaven reserved for machines which have a soul, was the Erector set. This toy was an assemblage of small steel beams and girders and 10,000 tiny nuts and bolts, all of which invariably rolled under the bed when you opened the box too quickly. The Erector set let you make elevators, cranes, bridges, roller coasters, and anything else little scientists and engineers could think of. There were no instructions. Just pictures of the finished product. You had to puzzle out which part went where, and in addition to developing hand-eye coordination, it forced you to develop pretty good spatial relations. Mostly, though, it was fun because not only did you play with it, but you had a sense of accomplishment when you finished a project.

Erector sets came in different sizes, and with the largest one, a number 12½, you could build a *robot!* This thing stood 4 feet tall, which was about as big as you were, it weighed 30 pounds, and it was all steel—no plastics. Instead of feet, it had caterpillar treads. All it could do was move forward—this was before computers, remember, but when it moved, it was *invincible.* You sort of pointed it, then plugged it in. Your brother, the cat, the dining room table; it didn't give a damn—it went right through all of 'em. The motor was geared down enough so that it could have pushed the Empire State Building off its foundation. Talk about a sense of power when you're 10.

The great and important thing about this Erector set was that it formed habits that are absolutely essential for a scientist—for that matter, for almost anybody who does complicated things. You *had* to have a fairly long attention span to make any-

thing worthwhile. The robot would take you at least a week to make, and you had to concentrate. That's exactly what I see here as a teacher of Bio 100. Most people who go down the drain here aren't dumb—they just can't concentrate very well, and their attention span is short.

Well, life for the young scientist in elementary school wasn't bad. It was actually pretty good. It goes without saying that we all wore glasses and were terrible in sports and that teachers loved us. This was not an unmixed blessing, because there were unpleasant consequences after school arising from a too cozy relationship with a teacher, but most of us could find some hulking friend who would be more than willing to say, "You calling my friend a brownnose?" in return for much needed assistance with homework. We were aware, like the political theorist Machiavelli, that in a reasonable world, power lies not in strength, but in knowledge.

Alas, with puberty comes that stage of life which is *not* reasonable and which respects form more than function. The lot of the youthful scientist in high school was not a cheerful one. Now it's computers, but in my time, the young scientists (who were not known as such, but rather by more colorful terms like "tool" or "geek") were relegated to student clubs like the Amateur Radio Association, which had tremendous prestige value with the cheerleaders. Today, fortunately, the genders suffer equally, but in the old days, the girls interested in science had it about an order of magnitude worse than did the boys. Ultimately, being a social outcast was beneficial, because by default, you spent a lot of time with books.

Although the social side of life in high school for the young scientist is apt to be a living hell, the academic side is usually a piece of cake. You don't have *that* much competition, unless you go to a very elite high school, and there are very few of those. There is thus a tendency to develop a sense of academic self-confidence that may be premature, as I was ruefully to discover.

Although I had a very good relationship with my parents, I wanted to go as far away from home as I could for college. I therefore found myself one fall appearing on the doorstep of a very prestigious eastern technical school, which shall remain nameless, at least in part because I start shaking uncontrollably every time I hear its name.

I had always thought they spoke English on the east coast, but being a good Pacific slope boy, I was unfamiliar with some of the accents, and it took me a while to figure out what everyone was saying. I finally found the all-male, all-freshman dorm and discovered to my absolute horror that freshmen were expected to take 19 credits their first semester; English, foreign language, calculus, physics, chemistry, and history. This was just a touch more work than I had been used to in high school.

College really was different in those days. Professors were concerned with neither the opinions nor the affection of their students. Students were some kind of annoying insect, always crawling around underfoot and making rasping, squealing noises when they were alarmed. There were no student evaluations of teaching, and the result was that the wildest kinds of eccentricities were tolerated and even applauded by connoisseurs of such things.

I still vividly remember the physics professor who loved to use these big demonstration devices to illustrate various principles of physics. To show how a gyroscope worked, he had this turntable he would stand on. Somebody would give him a little spin, and he could make himself spin faster or slower by extending or pulling in his arms. One day, he came to class so drunk that he forgot that you're supposed to extend your arms to go slower, and when he started to spin just a little too fast,

he clutched his arms to his body in terror. All we could see from the class was this colored blur on the stage, screaming "Gott in Himmel!"

Then there was a man I shall call "Professor Justice." He, too, was German, and very tidy and organized. On the first day of his chemistry class, he made a seating chart of the auditorium, and you had to sit in the seat you first sat in, all semester. He wanted an orderly exit, too, so at the end of the class, everybody in the back row had to stand up, then file out the back entrance. Then the next row would leave, until, finally, the front row could leave. Naturally, all the people in the front row were five minutes late to their next class. *Next* semester, everybody had figured out his system, so all the jocks occupied the back seats on the first day of class, and the rest of the class sorted itself out so the smallest and weakest got the front row. Professor Justice walked in, and said, "Gentlemen"—it was an all-male school with the exception of five women who had been admitted for the first time in my first year—"Gentlemen, those of you in the back row please rise. Now those of you in the front row rise. Now, exchange seats please." I always did like Professor Justice.

The rest of my first year was very much like your first year now, I would imagine. I fell madly in love with one of the five women, traveled, and made the awful discovery that the Boy Genius, when placed in genuine competition with 900 other boy geniuses, had some remarkable deficiencies in his background.

I remember those horrid, owl-eyed creatures who lived down the hall, who had domed foreheads and could look at a page for five seconds, then give it back to you word for word. We didn't have calculators then, but slide rules, a kind of me-

chanical ancestor of the calculator. Burt, two doors down from me, didn't need a slide rule. He would just look at a problem, blink his eyes a little, then write down the answer. It was *very* discouraging.

That first Christmas, I went home. They sent copies of your midterm grades to your parents in those days, so I knew that it would be impossible to come back as a conquering hero. My folks were very nice about it, very supportive, but I still knew I had to do *something* to get back the old swagger, the old self-confidence. I was a little late in the teenaged rebellion against parents, so I must have subconsciously decided that Christmas was going to be the time to do it. The first family dinner we had, aunts, uncles, grandparents, and other tribal kin, I really acted like a damned fool. Alternately surly and petulant, the nicer every one was to me, the worse I behaved. Profanity was never used in my family, so I decided that I would shock them to their very toes. At a lull in the conversation, I turned to my mother and said, "Pass the damned potatoes, please." Mom froze, stunned and appalled. Then my grandmother said exactly the right thing. She turned to my mother and said, "Don't just sit there, dearie, pass him the damned potatoes!"

Thus ended my teen rebellion.

I returned back east and fought off total disaster, but at the end of my freshman year, I had a 1.8 average and a profound and abiding loathing for chemistry, which was my major. I was forced to bid a tearful farewell to my one-in-five girlfriend, in which we pledged our undying love and promised to write each other every day. The train ride back home that June was the longest, saddest journey I've ever made.

After consultation with parents, friends, and conscience, it was decided that I should go to summer school at a nearby, also famous, school and see if I could pick up enough grade points so that I could transfer in as a regular student in the fall. To do that I would need A's in both my summer courses. I hadn't actually flunked out of the eastern school—I was as close to flunking as you could mathematically be without actually being there—but the whole thought of returning filled me with terror. It *had* to be A's.

Amazingly, as the summer went on, whether it was the climate, the challenge, or what, it looked as though I would get those A's. The final exams would tell the story. Live or die with the final. Through trial and error, by this time I had more or less evolved the study habits I told you about at the beginning of the semester. Don't fall behind; don't try to cram. Mental state and preparation were everything. The night before the final arrived. I stopped off at my mailbox at the dorm, and, sure enough, waiting for me was my daily letter from my eastern girl. I ripped it open with anticipation and read "Dear Stevie—This is the hardest letter I've ever had to write.... You know the rest of it. I was devastated, but I was also extremely fortunate because within minutes I discovered what great sages and philosophers had said for centuries. Out of adversity grows strength. I went up to my room and sat down 'til the shock wore off; then out of somewhere came this thought, "I *will* so become a scientist. I *will not* let this upset me. I *will* get those A's."

Next day, I went in and aced both finals.

The new school, and a new major, made all the difference in the world. It was still tough and demanding, but I had a much clearer idea of where I wanted to go. I'd always known I wanted to be a scientist, but I wasn't sure what kind. I'd always liked fooling around with frogs and toads as a kid, so I figured, "Go with what's interesting," and I became a biology major. I want to point out to you that this was not a *career* decision, it was an *interest* decision, which was much better. You can

get passionate and intense about an interest; a career is just the business part of your life. At the time, I didn't really know how I would make a living with biology, but I didn't care. Finally, I was studying something I could love.

There were a couple of teachers who had an enormous influence on me. In the best situation, you don't just learn facts and principles from a teacher—you learn values and attitudes, too. An excellent teacher has an enormous impact on people, which is why when dictators take over, the first people they arrest and kill are the teachers—they're too dangerous; they make you think.

I still remember Professor Gold, who taught developmental biology. He was a tall, homely man—homely like Abe Lincoln. We were all in awe of him. He never used notes for lecture, no matter how technical it was. The lectures were a dream to take notes from—organized, logical, and complete. When he made drawings, he'd use three pieces of chalk in each hand, to show tissue layers. His voice was powerful and mellow. He'd been a drama major as an undergraduate. Much later I got to know him as a colleague, and I found out that he had such terrible stage fright that, even after 20 years of teaching, he would go to the bathroom and throw up before every lecture. We never knew that as students. There are a lot of things you don't know as a student. In the laboratory, we'd all be peering into our microscopes, and he would silently come into the lab, stand behind you for a while, then ask, "What have you seen that's beautiful today?" I learned about professionalism and craftsmanship from Gold. Do a thing until you get it right, and don't turn in anything that isn't the very best that you know how to do. Good enough is not good enough. Attitudes and values.

We worshiped Gold, but we loved Professor Danetti. She was a herpetologist—reptiles and amphibians—which was a very unusual field for a woman in those days.

What was striking about her was the affection she showed for these curious creatures who have been condemned to spend their days on earth crawling on their bellies. One day, Danetti brought a box of sand and a bag into the lab. She cast the sand over the floor, then untied the end of the bag. A sidewinder rattlesnake slithered out. From our positions on top of the lab tables, where we had discreetly jumped when we saw what was in the bag, we could see the undulating patterns made by the snake as its ribs contacted the sand. When we had all seen the demonstration, Danetti walked over and started to reach for the snake. Somebody called out in alarm, "Dr. Danetti! Don't!" She smiled a soft smile, then said, "Oh, it's all right, they know when you mean them no harm," and gently picked up the rattler and put it in a terrarium. With my worldly sophistication and cynicism, I thought to myself, "Cute trick. Undoubtedly, it was defanged, so nothing to worry about." I thought that until the next demonstration when she showed us how the snake killed its prey. The mouse lasted about 10 seconds after it was struck.

Wonderful teachers like Gold and Danetti, and oddball eccentric teachers like the ones back east, provided both the seasoning and the substance of my undergraduate education. Most of the other instructors had as little effect as a raindrop on the ocean. But that is not very different from the rest of life. Excellence is a commodity not distinguished by its abundance.

When, by some miracle, I graduated, it was time to move to my real apprenticeship for science: graduate school. Grad school is very different from undergrad school. Its most striking characteristic is the personal relationship that you have between yourself and your "major professor." It is almost like a marriage, and lasts as long as some—five years or so.

I decided to go into the lab of Peter Bright, one of the most outstanding young scientists of his time. It was very exciting. Bright was a gadgeteer, as I was, and there was always some kind of crazy apparatus made of old coffee cans, scrapped electric razors, and burned out light bulbs sitting around. He loved to work in the field, which is probably where I acquired a taste for rough conditions. One time, I remember, we were in Australia, doing a project to measure the temperature under the skin of kangaroos. We had these tiny radio transmitters we were inserting under the skin. Didn't seem to bother the "roos," but you had to shoot them with a tranquilizer dart first. I knocked one down once, and Peter had its head cradled in his lap while he attached the radio. I guess I had the tranquilizer dose a little off, because the roo suddenly woke up, assessed the situation, decided he didn't like it, and bit poor Peter a stunning bite in the crotch. After, I'm happy to say, a complete recovery, Peter forgave me, but he decided to work on earthworms for a while after that.

The other grad students and I worked impossibly long hours, made virtually no money, and had no guarantee of a secure job when we finished, but we loved it. Like the old *Star Trek*, we were going where no one had gone before. We were working not for the rewards that the work brought, but for the reward that was the work itself. Science is not unique in this sense. Art, science, music, fine craftmanship of any sort are very similar. They all bring with them the feeling that you've *done something*. You may never be satisfied with your own work, but something has been accomplished.

Finally, I was finished. What had started with Mrs. Goldberg's cookies, progressed through baby brother's dread every time he heard big bro was going into the basement to work on the chemistry set, suffered near fatal derailment in that dismal freshman year, what had been rekindled by those saintly later teachers and fanned into flame by Peter Bright's infectious enthusiasm was now ready to be put to use. I was a scientist. Proud of it, and grateful for it.

And now it has been my turn to try to pass on to you those gifts which have been so generously given to me. I love being a scientist, standing at the edge of human-kind's knowledge of the universe, but I also love being a teacher. It is thrilling to think that I am teaching you things that I learned from my teachers, who in turn learned them from *their* teachers, in an unbroken chain going back thousands of years to the very first scientist-teachers. It is a sobering idea to think that a teacher is the guardian of one of our species' most precious possessions, the ability to transmit information from generation to generation, so that we may indeed learn from the past to improve our future.

So as I now look out at you, for the last time, I want to leave you with this. Follow your passion, whether it be science or some other demanding road. Do not be discouraged. If a kid with a 1.8 at the end of his freshman year can make it all the way, so can you.

As the semester now draws to a close, and you go to whatever destiny awaits you beyond these doors, let me say that it has been my privilege to be your teacher and I wish for you the same wonderful things it has been my great good fortune to experience. Good morning and...good-bye.

The spotlight is extinguished the instant he says his last word, leaving the auditorium in darkness. After a few seconds, the houselights slowly rise. The stage is empty, and Professor Farnsworth is gone.

Afterword from the Editor

Several months after Steve Farnsworth wrote the foreword to this collection of lectures, word was received from Indonesia that the plane carrying Steve and his field assistants had been mistakenly shot down with a Stinger missile by local rebels, and all aboard were feared lost. His friends back home were shocked but not surprised, and remembered that Steve had said that if he couldn't be shot by a jealous husband at the age of 81, he would prefer to die in the field, preferably on a rescue mission. We resigned ourselves to his loss, and although this book was not intended as a memorial, it seemed to have turned out that way.

Then a few weeks ago a returning Peace Corps volunteer passed through with a story that she had heard from another volunteer in the countryside in the Philippines. It seems that the Philippine government was investigating rumors about a white man who had floated ashore on a remote island in the Sulu archipelago and had set up a school, lavishly financed by the profits from a thriving smuggling business which brought Sony televisions into Mindanao *via* Borneo. When we heard this story, we all said, "That's Steve!"

So, Dr. Death, wherever, whoever, and however you are, this book's for you.

Frank Heppner
Kingston, Rhode Island

Acknowledgments

As you have discovered, this book is not a typical supplement to a biology textbook. For that reason, my first debt of thanks is to the corporate entity of McGraw-Hill, Inc., for its collective guts in taking a chance on something that was really different. Liz Dollinger, then at McGraw, got the project started, and shepherded it through the early days. Denise Schanck, Executive Editor of Science, College Division, McGraw-Hill, will be *Explanations'* midwife of record. One expects a biology editor to be competent and efficient, but one with a sense of humor is a jewel more precious than rubies.

The reviewers of this book had a particularly difficult time because they were charting new waters. *Explanations* is not a study guide and not a text, so how should it be reviewed? The editors at McGraw-Hill did a wonderful job of finding reviewers who truly understood the problems today's students have with the staggering amount of data in modern biology texts, and who were amenable to unconventional aids to these beleaguered scholars. I particularly want to thank Roy Scott and Steven Lawton of Ohio State, Linda Van Thiel and Lori Michalewicz of Wayne State, William Dunscombe of Union Community College, and Jean Heitz of Wisconsin, not just for their technical expertise, but for their tolerance of eccentricity. Good as they were, if any errors did manage to slip by, the responsibility is mine.

I can't forget Peg Barbour, who converted the unreadable to the polished, Dolores Chadwick, who got everything to New York, and the dozens of real-life, wonderful teachers who were the models for Farnsworth, but most of all I want to thank the students. When all is said and done, they're the ones who matter.

Appendix

Biology 100
Final Examination★
Fall Semester

Man is not a circle with a single centre; he is an ellipse with two foci.
Facts are one, ideas are the other.
—Victor Hugo

Name (print) _____
          Last   First   Initial

Name (signature) _____

Instructions: The correct answer is true in all its parts, and relevant to the question. There is only one correct answer to each question. There is no penalty for guessing. Print the answers on the answer sheet in block capital letters (A, B, C, D, E). Raise your hand if you have any questions, and a proctor will come to you. Pat the top of your head three times to indicate you have read and understood these questions. May the Force be with you.

1. *Life* is:
   A. A property held only by cells.
   B. A magazine.
   C. Defined by having the properties of responsiveness, metabolism, respiration, growth, and self-repair.
   D. Absent in things which don't reproduce.
   E. Ended when the heart stops beating.

2. Which of the following is *most likely* to be a polymer?
   A. p-o-i-u-y-t-r-r-e-w-q.
   B. 1-3-5-7-8-9-4-5-6.

★*Editor's note:* I have excerpted a small sample of the "concept" questions, based on the lectures, from Professor Farnsworth's final. Not included are questions of fact based on the textbook his students used. His final included 50 questions.

    C. 45-67-67-89-45-89.

    D. xc-e-we-de-w-ju.

    E. A stitch in time saves nine.

3. From a molecular standpoint, why do you die when you drown?

    A. Your blood is diluted.

    B. You switch over to anaerobic fermentation and are poisoned by too much ethyl alcohol.

    C. You are poisoned by a surplus of CO_2.

    D. You can't pick up the protons and electrons at the end of electron transport.

    E. The fact that you turn blue after you drown indicates the presence of the blue coloring material in water, which is toxic if it gets into the lungs.

4. The giant liver cell *Hepatica* speeds through hyperspace pursued by the barbaric Ethanolians. On board, however, there are problems which spell potential doom for captain and crew.

    "Gritty, give me more power!"

    "I can't, sir, I'm givin' ya all she can handle. There's enough reduced NAD^+ down here to phosphorylize half the ship."

    "Dammit Gritty, what's the problem?"

    "It's the _____ , sir. They've broken down completely!"

    A. Myochromes.

    B. Ribosomes.

    C. Cytochromes.

    D. Glycosomes.

    E. Dilithium crystals.

5. What's the main reason a plant looks green?

    A. Plants aren't actually green. They only appear that way because our eyes are more sensitive to green, and in *relation* to red and blue, the plant appears green.

    B. It absorbs green light, which makes the green pigment give off that color.

    C. It absorbs all colors, but green is given off as a fluorescence.

    D. It reflects red and blue, which are combined in our eyes to give the impression of green.

    E. It absorbs all the colors except green.

6. At a meeting of molecular biologists, it was discovered that there were two spies who were impersonating real molecular biologists. Sherlock Holmes was called in to detect the impostors. He first asked Helmut von Braun to spell *furry feline* in the DNA language. "Cytosine, Adenine, Thymine—CAT," replied von Braun. Holmes next asked John Ffolkes-Wombat, the Englishman, to spell *intestine* in DNA. "Sorry," replied Ffolkes-Wombat, "it can't be done." "Can you do it, Dr. Bangayan?" inquired Holmes. "Of course," replied the Filipino. "Guanine, Uracil, Thymine—GUT." Next he asked Dr. Blastoff if he could spell *to perform on stage* in DNA. "*Da*," replied the Russian. "Alanine, Cytosine, Thymine—ACT." "Monsieur la Barge," Holmes said, turning to the

Frenchman. "Would you be good enough to spell *to touch*, as in football, in DNA?" "*Tres simple*," responded la Barge. "Thymine, Adenine, Guanine—TAG."

After a moment, Holmes said, "Aha! The impostors are":
A. la Barge and Bangayan.
B. Blastoff and Bangayan.
C. Blastoff and von Braun.
D. von Braun and Ffolkes-Wombat.
E. Blastoff and Ffolkes-Wombat.

7. Which of the following would be the *best* example of semiconservatism?
   A. You build a model airplane from a set of plans.
   B. You take your negative to the photo store and have a couple of prints made.
   C. You take a salami, cut it in half, and give half to each of your friends.
   D. An amoeba divides, and after a while you have two new amoebas.
   E. You have your computer print out two copies of your term paper.

8. If the 2n number of chromosomes in an animal is 12, how many chromosomes will be in a cell in its liver?
   A. 6.
   B. You can't tell; it depends on what kind of animal it is.
   C. 24.
   D. 12.
   E. None; liver cells don't have chromosomes.

9. What is the probability of having *aabb* offspring if you cross an *AAbb* parent with an *aaBB* parent?
   A. 0 percent.
   B. 1 percent.
   C. 25 percent.
   D. 50 percent.
   E. 100 percent.

10. Which of the following communities would be *most likely* to be in Hardy-Weinberg equilibrium?
    A. Covertree—where all four of the natives live normal lives.
    B. Oldport—a remote paradise where the rich love to mingle with the richer, and hate the poor.
    C. Provenance—an area where nobody wants to go, the natives haven't the sense to leave, and only people with big ears have offspring.
    D. Cube Island—still another remote paradise, where the many natives indiscriminately mate with each other several times a day.
    E. Northerly—a far-distant place that endures an annual influx of people from Disconnecticut, who never leave.

11. Which of the following situations provides the best demonstration of a negative feedback loop?
    A. An Arkham student falls madly in love with a girl from Evans State. She tells him she'd rather go out with Pee Wee Herman, and he jumps off the

top of the Goforth Building, unfortunately landing on top of an accounting professor on her way to class.

B. A man walking down the street is accosted by a mugger. Unknown to the mugger, the man is a tenth-degree black belt in *tae kwon do* karate, and the mugger ends up with a ruptured spleen and all his teeth knocked out. The man laughs, takes the mugger's wallet, walks about 50 steps and drops dead. It later turns out his wife poisoned him.

C. Centipede Bus Lines drivers go out on strike and win substantial raises. The company then has to raise its fares, and passengers in droves leave for Acme Airlines. Centipede then lays off half its drivers, and the rest accept wage cutbacks to the original level to avoid further layoffs.

D. A lion comes up to a watering hole, where there is an antelope and a tourist from East Orange, New Jersey. The lion looks at both of them, then eats the tourist.

E. A student in Biology 100 goes berserk during an exam, closes his eyes, and circles random answers on the answer sheet. He gets 100 percent.

12. An example of an ecological population might be:
   A. All the fleas living on your dog.
   B. All the fleas living on your dog, your dog, and the grass your dog is standing on.
   C. You, your neighbor, and your neighbor's dog.
   D. All the animals living in the Delt house.
   E. The murderer, his victim, and the carrot used to stab the victim.

Answers

1. B	5. E	9. A
2. C	6. B	10. D
3. D	7. D	11. C
4. C	8. D	12. A

Index

*This index represents a listing of important terms at their first occurrence in the text.